COSMIC COUNTDOWN

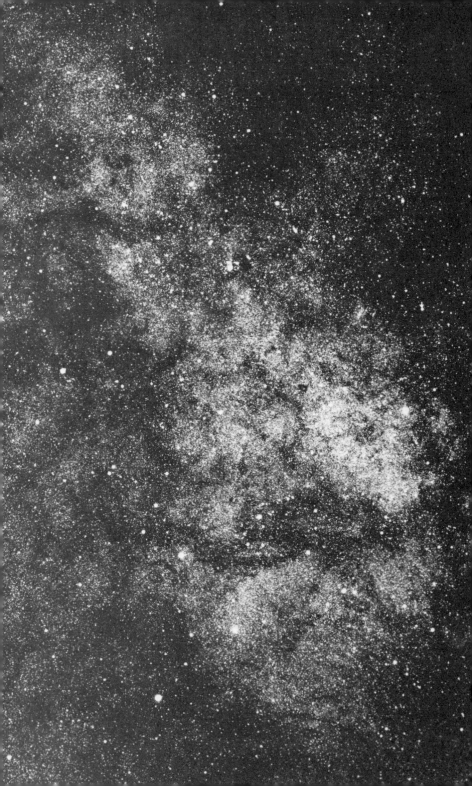

COSMIC COUNTDOWN

WHAT ASTRONOMERS HAVE LEARNED ABOUT THE LIFE OF THE UNIVERSE

by FRANCINE JACOBS

Foreword by Dr. Jastrow
based on works by Dr. Robert Jastrow

M. EVANS AND COMPANY, INC.
NEW YORK

Library of Congress Cataloging in Publication Data

Jacobs, Francine.
  Cosmic countdown.

  Includes index.
  Summary: Examines the new discoveries and revolu-
tionary ideas that have transformed the ancient study
of the stars into the complex science of astronomy.
Also discusses theories about the beginning of the
universe and what is likely to happen to it in the
future.
  1. Cosmology—Juvenile literature. 2. Astronomy—
Juvenile literature. [1. Cosmology. 2. Astronomy]
I. Jastrow, Robert, 1925-   II. Title.
QB981.J235 1983   523   83-5535
ISBN 0-87131-404-5

M. Evans and Company, Inc.
216 East 49 Street
New York, New York 10017

Design by Lauren Dong

Manufactured in the United States of America

9 8 7 6 5 4 3 2 1

(frontispiece)
THE MILKY WAY GALAXY. Our sun, a medium-sized star, is
but one of some 200 billion stars that make up the Milky Way
Galaxy. Only a few thousand of these stars can be seen on a
clear night with the unaided eye. Telescopes bring many more
stars into view. This photograph of our Galaxy was taken
through a 10-inch telescope; the film was exposed for several
hours to record the faint light of more distant stars. About
10,000 stars are visible here, even though the picture shows only a
thousandth of the area of the full night sky. (Yerkes Observatory)

for Fran and Wally Herbert

Contents

Foreword

Astronomy is the oldest science and the one with the greatest fascination for everyone. At one time it was a field that changed very little from year to year, and the great work was done by a small band of dedicated individuals, peering through their telescopes and pondering the mysteries of the Cosmos as revealed to them in the night sky.

Today, all that is changed. New astronomical discoveries come almost daily; the Universe is filled with surprises, wracked by titanic explosions, populated by exotic objects like pulsars, quasars, and black holes. Most remarkable among all these discoveries is the new evidence relating to the beginning of the world.

Astronomers have been able to prove, by the most subtle and ingenious methods, that the universe had an abrupt beginning. According to the astronomical evidence, our Universe—the World in the Biblical sense—came into being, somewhat as the Bible said it did, in a flash of light

9

and heat, about 15 billion years ago. In my opinion, this discovery contributes more to our understanding of our place in the Cosmos than any other finding in science.

Especially illuminating is that figure of 15 billion years—the astronomer's value for the age of the Cosmos. We know from the ages of moon rocks that the solar system, including the Earth and all life on the Earth, is only 4.6 billion years old. Consider the meaning of those two numbers, 4.6 billion years and 15 billion years. They tell us that most life in the Universe—if it exists—is likely to be billions of years older than we are. That is an extraordinary finding, because a billion years ago the highest form of life on the Earth was a worm.

In other words, the "average intelligent person" in the Universe is as far beyond the human being as the human being is beyond the worms in the garden. Although people are the wisest and most intelligent creatures on this particular planet, relative to the cosmic scale of time their brain has not had very long to develop. Humans are recent arrivals in the world of intelligent life—the youngsters, the new kids on the block. The life on other worlds will be nothing like the lovable E.T., with his vaguely human form, or the funny creatures standing at the bar in *Star Wars*. Our brothers and sisters of the Universe, if they exist, will have bodies completely different from anything we can imagine, and scientific powers that will seem to us to be pure magic.

Astronomy has another message to communicate regarding the origins of human beings and their place in the Cosmos. According to the story of the red giants and the white dwarfs, and the modern theory of cosmic evolution, every rock, every plant and flower, and every animal and human being on the face of the Earth is made out of atoms that were created in stars many billions of years ago and then dispersed to space when those stars exploded and died. This is the message of modern astronomy: The atoms in our body were made in stars; we are the children of the stars.

Astronomy has something to say about our future as well as our past. It predicts that in 6 billion years the sun will become a red giant. At that time, every living thing on the Earth will be destroyed by the heat of our dying star. But that will not happen for billions of years. Before then, our offspring will have fled to another planet, perhaps in another galaxy.

These are some of the exciting prospects opened up by the new discoveries in astronomy. Unlike the work of science fiction writers, they rest on solid scientific findings regarding the evolution of the Universe and the birth and death of stars and planets.

I first learned about the new discoveries through my work in the National Aeronautics and Space Administration (NASA), the space agency, where I found myself talking to some of the world's greatest astronomers and planetary scientists. Their ideas seemed so exciting to me that I decided to put them into a series of books written in clear language without a great deal of mathematics, so that everyone could understand them and share their excitement with me. Three of these books, *Red Giants and White Dwarfs, Until the Sun Dies,* and *God and the Astronomers,* have appeared. I have also written a textbook, *Astronomy: Fundamentals and Frontiers,* with Malcolm Thompson. Francine Jacobs has taken the astronomical sections from my writings and combined them into a single narrative of her own without sacrificing accuracy or omitting any important ideas. Through her talent as a writer, the great discoveries of astronomy and the Space Age are now accessible to every young reader in a clear, up-to-date and authoritative form.

Dr. Robert Jastrow
Founder, Goddard Institute for Space Studies
    of the National Aeronautics and Space
    Administration
Professor of Astronomy and Geology,
    Columbia University
Professor of Earth Science,
    Dartmouth College

The stars, planets, moons, and bits of cosmic dust scattered in the great vastness of space, our Earth, its mountains and oceans, animals, people and plants, the tiniest bacteria, atoms, and their smallest parts—all matter throughout space and time—make up the Universe, the Cosmos.

Prologue

The sun will die 6 billion years from now. But long before then the Earth's inhabitants will probably have escaped to another planet in the Milky Way Galaxy.

Imagine the Earth in 200 million years, as far in the future as the early dinosaurs are in the past. The world is different. Gone are the masses of ice that once covered the North and South poles. The seas have risen and flooded large areas that once were coastal plains.

A spacecraft stands ready for lift-off on a launching pad up on Grand Mesa, a 2-mile-high plateau, in western Colorado. The huge machine looks like a great, winged reptile. It is before dawn, yet even in this high place the temperature is already warm. Now, slowly a pink glow brightens the horizon, announcing the new day. The sun appears. It is brilliant and beautiful. But the sun is slowly growing and will destroy the life it once nurtured on our Earth. The sun is becoming a great, shining monster, a swollen red giant, that is drawing our small planet closer and will one day consume it.

Aboard the spaceship, astronauts complete their final preparations and settle back to wait for lift-off. These men and women will work and live together for many years. Their mission is to investigate a large planet in another part of the Galaxy, a warm planet that might have water and an atmosphere similar to that of Earth. To accomplish this, they will travel beyond our solar system. Powered by a new form of high-energy fuel, the spaceship will reach a speed of more than 160,000 miles per second, approaching the speed of light (approximately 186,281 miles per second). Far out in space the astronauts will seek a new home for the peoples of Earth before the sun dies.

COSMIC COUNTDOWN

1

The Changing Cosmos

Less than 100 years ago, people still believed that the
Cosmos consisted of one great cluster of stars, one
galaxy. Our solar system, with its spinning, sun-orbiting
planets, was but one small part of this Milky Way Galaxy.
The distant stars and the rest of the Cosmos appeared to
be motionless and unchanging.

There were some astronomers who suspected that
certain luminous, cloudlike objects in the heavens, nebulas,
might be other, separate galaxies. But photographs of
these glowing bodies did not reveal how distant they were.
Pictures of objects in space taken through a telescope do
not tell how far away they are because a body that is
enormous in size and very bright may look small and faint
if it is very distant. Many astronomers argued that the
strange, glowing nebulas were relatively small pockets of
luminous gas swirling nearby in the spaces between the
stars.

VESTO MELVIN SLIPHER (1875–1969). *In the tradition of many great scientists, Vesto Slipher did not disregard an accidental discovery but investigated it further. At the time he reported his discovery, he did not understand the full significance of his finding—that neighbor galaxies were apparently racing through space away from us. This modest American astronomer's contribution led to an entirely new concept of the Cosmos.*

Then, in 1912, an American astronomer named Vesto Melvin Slipher made an accidental discovery that was to change forever our ideas about the Cosmos. Slipher worked at the Lowell Observatory in Flagstaff, Arizona. The observatory, a building with a domed roof that opened to admit starlight, housed a modest, 24-inch telescope that Slipher was using to investigate one of these mysterious glowing clouds called the Andromeda Nebula.

The spiral shape of Andromeda suggested that it might be spinning, but it was too distant for Slipher to be able to observe this. So he tried an indirect method to determine if the nebula was in motion. He studied the light that came from Andromeda. Light is a form of energy that travels in waves from its source. Physicists had discovered that motion affects the length of light waves. Light waves from an object that is moving toward us become bunched and shortened. We see this light as the color blue. If, however, a glowing object is retreating, moving away from us, its light waves appear as the color red, and this is called the *red shift*.

Slipher took advantage of this phenomenon to see if he could detect shifts in the light waves from Andromeda that would prove it was rotating. He inserted a prism into his telescope. The prism, a triangular piece of glass that separates white light into its component colors, violet, blue, green, yellow, orange, and red, enabled Slipher to analyze the makeup of the light from the nebula and to compare it to light from a stationary source.

The result of this test confounded the 37-year-old astronomer. He could not tell whether Andromeda was spinning, but the strange, glowing nebula was certainly moving. Light waves from Andromeda showed a dramatic shift. If his calculations were correct, Andromeda was traveling through space at the stunning speed of 700,000 miles an hour.

Slipher was surprised by his accidental discovery. He was curious, too. Were the other nebulas also moving through space? He used the same technique to study them. He photographed more than a dozen through prisms over the course of the following year. To his

amazement, nearly all the nebulas he could see through his relatively small telescope were indeed rushing through space, some at incredible speeds of up to 2 million miles an hour. In almost every case these nebulas produced a red shift—they were moving *away* from the Earth.

Slipher had stumbled onto something extraordinary. But his findings puzzled him. What did they mean? Why were these mysterious nebulas flying through space? Why were they rushing away from us? According to the laws of chance, one might have expected that some would be moving toward us and others away. But this was not the case. Did this suggest that the whole Cosmos was moving away from one special point in space, and that the Earth was at this point?

Slipher assembled his notes, gathered his slides—proof that the nebulas were speeding through space away from us—and modestly presented his findings to a meeting of the American Astronomical Society in 1914. When he had finished his presentation, all 76 of his colleagues at the meeting rose and cheered in an unusual gesture of ·admiration. Though no one yet understood the full meaning of Slipher's work, they recognized that he had made an important discovery. Accidentally, Slipher had hit upon

THE ANDROMEDA GALAXY. *Once called a nebula, a luminous cloud, Andromeda is a true galaxy composed of stars. It is similar in size to our own Milky Way. In this photograph, the bright center of this spiral-shaped galaxy is an extremely crowded cluster of stars, so dense that their lights cannot be seen separately even through the largest telescopes. On either side of Andromeda the two prominent bright spots represent satellite galaxies each containing several billion stars. (Lick Observatory)*

evidence that the Cosmos is neither motionless nor unchanging. He had found the first clue that the Cosmos is expanding.

Slipher returned to the Lowell Observatory to continue his work. Meanwhile, across the Atlantic Ocean, another discovery of great importance was made by the famous German-born physicist, Albert Einstein. Working in Berlin in 1916, Einstein conceived a mathematical theory concerning gravity, the attractive force exerted by one mass on another. Einstein described the mathematical relationship between the mass of an object and its gravity. He further predicted the influence of gravity on the course of bodies in motion. His formulations gave astronomers important clues to the character and shape of the Cosmos.

A Dutch astronomer, Willem de Sitter, recognized that Einstein's ideas meant that the Cosmos is expanding. Einstein had not foreseen that his theory predicted this. He refused to accept de Sitter's conclusion. The notion that the Cosmos is expanding suggested that there was a time when the Cosmos must have been all one and to-

ALBERT EINSTEIN (1879–1955). Considered handicapped by learning disabilities as a schoolboy in Germany, physicist Albert Einstein conceived and published a radical new mathematical explanation of gravity when he was 37. This general theory of relativity revised centuries-old laws of physics and became the foundation for modern physics and astronomy. A warm, pleasant man, Einstein was driven from his homeland in the 1930s by antisemitism. He became an American citizen in 1940, and spent most of his remaining years pursuing his research at Princeton University in New Jersey (Copyright Hebrew University of Jerusalem)

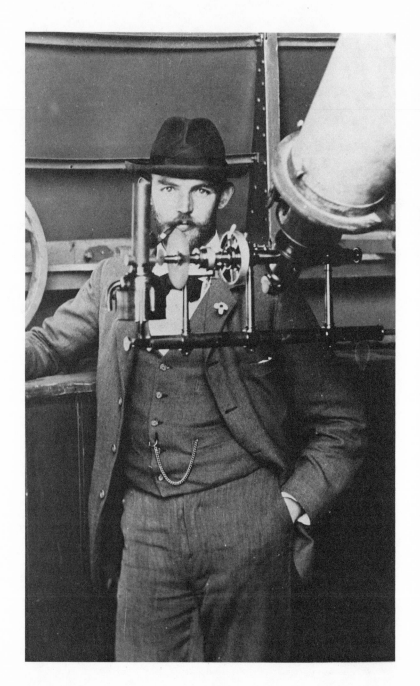

gether. This meant that the Cosmos had not existed forever; it had a beginning. Might it also have an ending? Einstein found such ideas very disturbing, and he rejected them.

While these theories were debated in Europe, the First World War (1914–1918) was taking place. The war interrupted normal communications between scientists in Europe and the United States. De Sitter was unaware that Slipher had unknowingly come upon evidence to support his theory that the Cosmos is expanding. But after the war, an American astronomer named Edwin Powell Hubble learned of de Sitter's theories and was impressed with them. Hubble had been in the audience that had heard and applauded Slipher's presentation of evidence that the nebulas were moving away from the Earth.

Edwin Hubble was probably the first American astronomer to understand the connection between Slipher's discovery that the nebulas are moving rapidly apart and de Sitter's theory that the Cosmos is expanding. In the mid-1920s, he set out to prove what Einstein was reluctant to accept.

WILLEM DE SITTER (1872–1934). *In 1916, de Sitter, a Dutch astronomer working in Leiden, Holland, received a copy of Albert Einstein's paper on the general theory of relativity from the physicist, who was then working in Berlin. De Sitter recognized a startling implication in this theory. According to Einstein's equations, the galaxies of the Cosmos must be moving rapidly apart. De Sitter played an important role in establishing the revolutionary concept that the Universe is expanding. (Martin Smit)*

Hubble had an unusual background for a scientist. He was not only a scholar but also a fine athlete, a championship basketball player, and a boxer. He studied physics, briefly practiced law, and was wounded in action in the First World War before turning to astronomy. Hubble teamed up with Milton Humason at the Mount Wilson Observatory near Pasadena, California. Humason also had an unusual background for an astronomer. He had started out as a mule-train driver and become a janitor at the observatory. There, he taught himself astronomy. In fact, he had developed extraordinary skill in handling the observatory's reflecting telescope, the largest telescope in the world at that time.

A reflecting telescope uses a bowl-shaped mirror to gather and focus light. It differs from the refracting telescope—the kind that Slipher had used—which employs a system of glass lenses to concentrate light. The size of a telescope refers to the width of the mirror in a reflecting telescope, or the main lens in a refracting one.

The huge reflecting telescope at Mount Wilson has a 100-inch mirror. Painstakingly, night after night, in the chilly cool of the mountaintop, Humason searched space for nebulas and measured their speeds and direction as Slipher had done. But with the advantage of the huge

EDWIN POWELL HUBBLE (1889–1953). Hubble (on the right) with colleagues James Jeans and Walter Adams, at the Mount Wilson Observatory, Pasadena, California. Hubble set out to determine if the Cosmos is expanding by accurately measuring the distances to nearby galaxies. (American Institute of Physics, Niels Bohr Library)

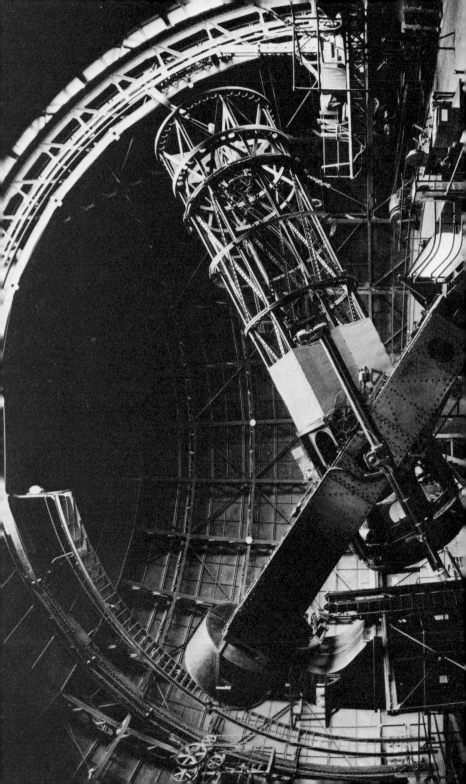

telescope he could see nebulas that had been too faint and too distant for Slipher to find with his 24-inch instrument. Humason's observations confirmed Slipher's findings. All the nebulas he discovered were hurtling away from the Earth and at exceedingly high speeds. Some were traveling at the incredible rate of 100 million miles an hour.

While Humason patiently searched for new nebulas, Hubble photographed the ones that Slipher had studied. He discovered that the nebulas were not glowing clouds of luminous gas at all. Nor were they part of our own Milky Way Galaxy. They are true galaxies themselves, billions of stars grouped together in giant clusters, separated from our own Galaxy and from others by vast distances. Hubble took the small size of these spiral-shaped galaxies as they appeared in the 100-inch telescope and the faintness of their light to be an indication of just how great the distances between the galaxies must be. By comparing the faint light of distant stars to the brightness of a star in our own Galaxy, he could gauge how far away they were.

THE MOUNT WILSON TELESCOPE. This huge reflecting telescope was the largest in the world in the 1920s. The telescope's 100-inch-wide mirror gathers light from sources up to 100 million light-years away (600 trillion miles). Photographs taken through this instrument revealed that mysterious spiral nebulas, such as Andromeda, contained billions of individual stars. The Mount Wilson telescope helped to establish that the nebulas are true galaxies, separate island universes. (Mount Wilson and Las Campanas Observatories, Carnegie Institution of Washington. Mount Wilson Observatory photograph)

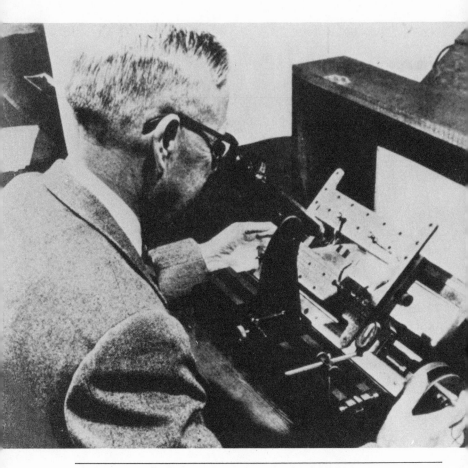

MILTON HUMASON (1891–). *Humason measured the speeds of the retreating galaxies. His method, first used by Vesto Slipher, depended on an effect known as the red shift. Light waves from an object that is moving away are "stretched," or lengthened; they redden in color. The degree of reddening, or red shift, is a measure of the object's speed. In this photograph, Humason compares the red shift of a distant galaxy, recorded through the telescope on a half-inch slip of glass, to light from a nonmoving source.* (Scientific American)

Hubble calculated the distances to about a dozen nearby galaxies. To do this, he used a yardstick called the light-year. A light-year measures distance, not time. It is the distance that a ray of light covers in one year.

As we have said, light moves at about 186,000 miles per second:

Multiply this by 60 for the seconds in a minute.
Multiply by 60 again for the minutes in an hour.
Multiply by 24 for the hours in a day.
Multiply by 365¼ for the days in a year.

Now you begin to appreciate how many miles a light-year represents. A light-year is 6 trillion miles. Most of the galaxies that Hubble measured were more than a million light-years away, and the farthest of them was 7 million light-years distant. These distances are greater by far than the size of our Galaxy, which is 100,000 light-years across. Hubble was the first to give us an inkling of the vast size of the Cosmos.

With the distances to Slipher's and Humason's galaxies in hand, Hubble was curious to see whether there was a relationship between the distances of the galaxies and the speeds at which they were moving. So he plotted on paper distance against speed to form a graph. This revealed a clear relationship: The farther away a galaxy is, the faster it moves. The Cosmos was not only expanding, but it was moving outward evenly in all directions. This extraordinary discovery applies not only to the Cosmos but to all uniformly expanding bodies. It is now known as Hubble's Law.

To illustrate Hubble's Law, let us use a Chinese checkerboard to represent the Cosmos. Place a marble, to serve as a nearby galaxy, 2 spaces from the center. Place another marble, to stand for a distant galaxy, 6 spaces from the center. If we could expand the checkerboard, doubling its size, we would find that the marble, representing the nearby galaxy, would now be 4 spaces away from the center. But the marble serving as the distant galaxy would now be 12 spaces away. In the time it took to expand the checkerboard, the marble nearest the center moved only 2 spaces, while, in the same time, the marble representing the distant galaxy traveled 6 spaces. Clearly, the marble standing for the distant galaxy had traveled faster.

Hubble's observations and measurements established that the Cosmos is expanding in all directions away from the Earth. Did this mean that our planet is the center of the Cosmos? The idea of an Earth-centered world had been accepted as fact up until the sixteenth century. Then the Polish clergyman and astronomer, Nicholas Copernicus, convinced other observers that the Earth and its neighbor planets revolve around the sun; the Earth could not be the center of the Cosmos.

Was Copernicus wrong? No. For if we were to view the galaxies from any other place in the Cosmos they would appear to be moving away from that place also. The Cosmos has no center. All the galaxies are simply moving outward and away from one another.

Teachers of astronomy often compare the Cosmos to a balloon with dots on its surface, representing the galaxies. If the balloon is filled with air, all the dots will move apart and away from one another. No one dot can

SPEED
IN MILES PER HOUR

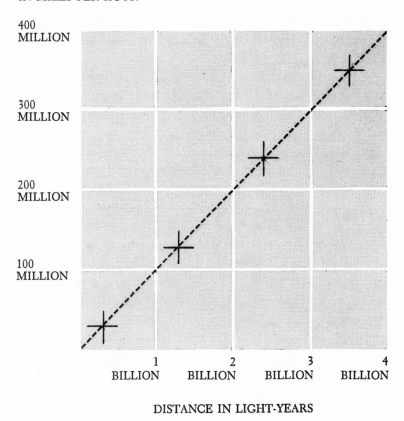

HUBBLE'S LAW. *The line indicates a simple proportion between speed and distance. If one galaxy is twice as far away from us as another, it will be moving twice as fast; if it is three times as far away, it will be moving three times as fast, and so on. This proportion is the mathematical statement of Hubble's Law.* (God and the Astronomers *by Robert Jastrow, published by* Warner Books)

be considered to be at the center. Another well-known example compares the Cosmos to a cake batter with raisins in it. When the batter is baked and the batter rises, the raisins move outward and away from one another throughout the cake.

So often in science the solution to a problem seems simple once the answer is discovered. It is the process of discovery that is so complicated. Slipher stumbled onto the finding that the galaxies are retreating, and that in turn provided the crucial step that directed Hubble and others onto the path toward a major advance in our understanding of the Cosmos.

In the space of two decades, beginning just before the First World War, age-old ideas were suddenly swept away by revolutionary discoveries. The Cosmos is neither motionless nor unchanging, it is much more vast than anyone had imagined, and it is expanding at a rapid rate. Why? What does this mean? What does it tell us about the origin of the Cosmos? Was there a beginning, a creation?

2

CREATION: THE BIG BANG

IMAGINE for a moment that we have just entered a darkened movie theater. Up on the huge screen before us, a motion picture of the Cosmos is in progress. Great galaxies with billions of stars are hurtling apart through space. We are puzzled by what we see. So we ask the projectionist to run the film backward. Now all the galaxies on the screen retreat toward one another until they join. We realize now that we have been watching the effects of a great cosmic explosion.

Is this what is happening in reality? Did such a cosmic explosion occur in the past to account for the galaxies flying apart? Is it possible that the Cosmos is blowing up before our very eyes?

In 1930, Albert Einstein left Berlin and traveled to southern California to the Mount Wilson Observatory to see Hubble. He studied Hubble's slides and looked in his telescope. The visit convinced Einstein that indeed the

Cosmos is expanding, and rapidly. De Sitter was right—the Cosmos is exploding.

The following year, in 1931, a Belgian astronomer, a Jesuit priest named Georges Lemaître, conceived the idea that the whole Cosmos was once compacted together into one mass. In his mind's eye he saw this body as an incredibly dense droplet containing all the matter and energy that was to be our Cosmos. It was a mass with fantastic internal pressures and a temperature trillions of degrees hot; it was a cosmic egg ready to hatch. When it did, at some critical moment, an explosion of unimaginable power —a blast infinitely greater than the explosion of the most powerful hydrogen bomb—blew the mass apart, creating the Cosmos.

Lemaître's ideas were widely discussed among scientists. A few years later, in 1936, physicist George Gamow, who had emigrated from Russia to the United States, added to Lemaître's theory. Gamow suggested that all the chemical elements in the Cosmos—the material out of which all matter is formed—owe their origins to the catastrophic event that Lemaître had described. Gamow called this theory of the explosive birth of the world the Big Bang theory.

Astronomers have assembled some convincing evidence in support of the Big Bang. They have measured the distances between the galaxies and the speeds at which they are moving away from one another. And they have shown, by tracking the galaxies backward, that they can be traced to a common origin. The astronomers, in effect, have accomplished mathematically what the projectionist in the example earlier did by running the movie of the cosmic explosion in reverse. According to their calculations, all the galaxies were together some 15 billion years ago.

The Big Bang theory forces us to give up some of our most cherished ideas. It deprives us of the comforting notion that the Cosmos always existed. It says instead that the Cosmos had a definite beginning. Science and religion agree on this important point: There was a creation. In fact, the biblical and scientific accounts of this event are similar. Both versions describe creation as happening at one precise moment, in a great, blinding flash of light.

Try to imagine the Cosmos at birth. It is a formless world of intense, radiant energy, a dazzling, white fireball at 10 billion degrees. Within a few minutes, the expanding new world cools to 1 billion degrees. Infinitely tiny, scattered bits of matter form now. Over the next millions of years, gravity gathers many of these particles together, and simple atoms of hydrogen are created. The brilliant light and heat of the newborn Cosmos begins to fade over this time. A soft, glowing cloud of cooling hydrogen gas and cosmic dust remains.

A billion years pass. The cooled, luminous hydrogen cloud dims further and contracts to form the galaxies. The Cosmos is dark now, but within the galaxies brilliant stars, like shining jewels, are born. Over time stars die but others appear. This process, the birth and death of stars, continues. In the Milky Way Galaxy, our sun is born 4.6 billion years ago. Other matter collects near this star and is captured by the sun's gravity to form our solar system. These smaller masses that circle the sun and shine by its reflected light are planets; one of them is our Earth.

THE BIG BANG THEORY. *Astronomers theorize that 15–20 billion years ago the Cosmos was a highly compressed cloud of matter and energy that suddenly expanded outward with great violence. Figure 1 represents the Cosmos 1 second after the explosion that marked its birth. At this time the temperature of the Cosmos was 10 billion degrees. In the second figure, 3 minutes later, the Cosmos continues to expand rapidly, but the temperature has dropped to 1 billion degrees. Infinitely tiny bits of gaseous matter are combining to form larger particles. In figure 3, a billion years have passed and the gases of the Cosmos are considerably cooler. Atoms have formed and begun to condense into galaxies and into stars and planets within each galaxy. The final figure shows how the expanding Cosmos would appear from any one of these galaxies. Each galaxy would seem to be a center from which all other galaxies are retreating.* (Adapted from Atlas of the Universe, *published by Elsevier/Nelson*)

1   2

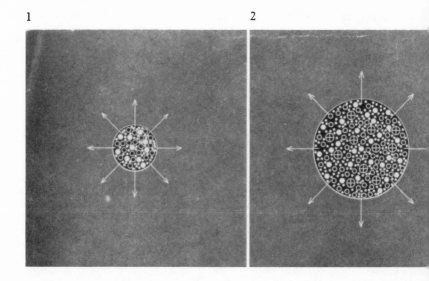

In time a wondrous process called life begins on the Earth and eventually people appear.

Since the beginning of the world, about 15 billion years ago, there have been great changes in the Cosmos. But what do we understand when we think of 15 billion years? A billion is 1000 million, and trying to grasp the meaning of a million years is enough to boggle our minds. Such vast amounts of time are beyond understanding. It does help if we reduce time to some scale with which we are familiar.

Let us, therefore, invent a cosmic clock. We will compress the 15-billion-year life of the Cosmos to one full day on this clock. A billion years will be roughly 1 hour. 10 million years will be 1 minute. The whole span

3   4

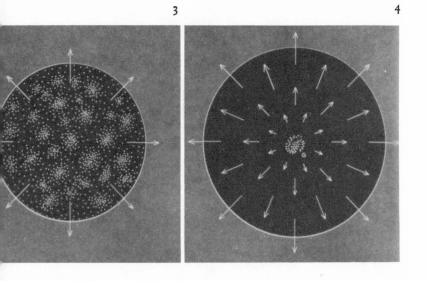

TODAY

2 MILLION YEARS AGO — — THE FIRST HUMAN APPEARS
(HOMO ERECTUS)

30 MILLION YEARS AGO — — THE MONKEYS AND APES EVOLVE

65 MILLION YEARS AGO — — THE DINOSAURS DISAPPEAR
THE MAMMALS INHERIT THE EAR[']

200 MILLION YEARS AGO — — THE FIRST DINOSAURS
THE FIRST MAMMALS

300 MILLION YEARS AGO — — THE FIRST REPTILES

450 MILLION YEARS AGO — — THE FIRST FISHES

600 MILLION YEARS AGO — — THE FIRST HARD-BODIED ANIMAL[S]

1 BILLION YEARS AGO — — MANY-CELLED ANIMALS

3.5 TO 4
BILLION YEARS AGO — LIFE APPEARS
BACTERIA AND SIMPLE PLANTS

4.6 BILLION YEARS AGO — — THE SUN AND EARTH APPEAR

15-BILLION-YEAR GAP —

— STARS AND PLANETS FORM
THROUGHOUT

1 BILLION YEARS LATER — — STARS BEGIN TO FORM

20 BILLION YEARS AGO — THE UNIVERSE EXPLODES
INTO BEING

of human civilization is less than a tick of our clock—one-thirtieth of a second.

Time begins at midnight when creation takes place. Thirty minutes later the galaxies, stars, and planets begin to appear. They will continue to form all through the night and the day to follow. The next evening, at 6 P.M., our solar system forms—the sun, the Earth, our moon, and the other planets appear. Near day's end, at 5 minutes to midnight, living creatures first crawl out of the sea onto the land. Four minutes later the dinosaurs rule the Earth. They disappear in less than a minute. Modern human beings step out on the scene exactly 1 second before midnight.

We have traced the chain of events that led from the Big Bang, the moment of creation, to intelligent life on Earth. But what caused creation in the first place? Why did it happen? Was it the hand of God? Science cannot say. The birth of the Cosmos defies the known laws of physics—these principles fail to explain what force caused creation. That fearsome power remains a mystery to science.

And what came before creation? Was there an earlier Cosmos? Did it contain all the energy and matter of our Cosmos? Was it a Cosmos similar to our own, or one very different? What kind of history did it have? It is almost

---

THE COLUMN OF TIME. *The great events in the history of the Earth and life on the Earth are shown here. The bottom of the column represents the moment of creation. Whether the creation of the Cosmos happened 15 or 20 billion years ago is uncertain. The important point is not precisely when the cosmic explosion occurred, but that it happened at one precise moment many billions of years ago.*

certain that we shall never know. For if an earlier Cosmos existed, it is not likely that any clue to it survived its destruction. The evidence would have been melted down and consumed in the fiery explosion that created our world.

All we can be certain of is that creation took place. The Cosmos had a beginning.

3

COSMIC CONTROVERSY

I HAVE no axe to grind in this discussion [but] the notion of a beginning is repugant to me. . . . I simply do not believe that the present order of things started off with a bang . . . it leaves me cold," declared the distinguished British astronomer, Arthur Eddington.

The Big Bang theory was not accepted immediately. It stirred debate and controversy in the scientific community. The evidence that the Cosmos had a definite beginning was challenged. Albert Einstein had difficulty accepting it. For it is hard to conceive that the Cosmos was not always here. This leads to the troublesome question, "What came before the beginning?" Even Saint Augustine, pondering the wonder of creation, found himself asking what God was busy with before He made Heaven and Earth. And finding no other answer replied, "He was creating Hell for people who asked questions like that."

Difficult as it is to conceive of a beginning, it is even harder to accept that there will be an end. We want to reject the notion that the Cosmos will not always be here. But the evidence that the Cosmos had a beginning also leads to the conclusion that the Cosmos will have an end. Its violent, explosive birth began a process that will go on to the end of time, when the Cosmos dies. The galaxies will continue to rush apart and away from one another, and the space between them will increase and grow emptier. Over time, the stars that make up the galaxies will grow old. They will burn up their hydrogen gas and die. New stars that arise will also age and eventually burn down, too. Finally, when the glow from the last star winks out, all light will vanish from the Cosmos. The cosmic countdown will be complete. The Cosmos will have run its course, and its life will end.

Is there no other answer, no explanation without such a doomsday, to account for the changing Cosmos? Any scientific theory that attempts to explain the Cosmos and predict its fate must deal with the key fact that it is expanding. There is no other way.

The evidence that the Cosmos is expanding, however, did not automatically discourage all those who rejected the Big Bang theory. Some astronomers, in 1948, still stubbornly insisted that there was no beginning and there could be no end. They held that the Cosmos was in a steady state, that it had always existed, and always would. According to this theory, new matter is continually forming—it arises from nothing. This new matter maintains the Cosmos in a steady state that will endure forever.

George Gamow, one of the originators of the Big Bang concept, led the argument against the steady-state theory. Good-humored, but also tough, Gamow was a convincing spokesperson and the author of many popular books on astronomy. In 1948, Gamow and his associates, physicists Ralph Alpher and Robert Herman, claimed that proof of the Big Bang theory existed and would be found in the Cosmos. The crucial evidence would be in the form of radiation, a faint, leftover glow from the white-hot inferno of creation.

But confirmation of the afterglow's existence, the proof essential to settle the debate over the Big Bang theory, did not come immediately. Though extraordinary improvements had been made in telescopes, cameras, and other optical equipment, these systems sensed only visible light. The afterglow that the astronomers were searching for was a form of radiation that was invisible. Instruments sensitive enough to detect the lingering radiation already existed, but scientists familiar with these devices either did not know about the ideas of Alpher and Herman, or did not take them seriously.

Fortunately in science, however, it frequently happens that an apparently unrelated discovery provides the means to solve a problem. This proved to be the case in the search for the afterglow of the Big Bang. In 1931, a researcher named Karl Jansky at the Bell Telephone Laboratories in Holmdel, New Jersey was trying to solve a problem in transatlantic telephone communications. Conversations between London and New York were frequently interfered with by static and hissing sounds. Jansky looked for the source of this noise. He found that it came from outer space, from the center of the Milky Way Galaxy. Jansky had accidentally discovered the existence of radio waves in space. Radio waves, which we produce here on

Earth for communication, are also generated in the Cosmos. Astronomers began constructing radio telescopes (huge dish-shaped antennas) to catch these emissions.

In 1965, Robert Wilson and Arno Penzias, two physicists also at the Bell Telephone Laboratories in New Jersey, were working with a great, horn-shaped radio telescope on a hillside in Holmdel. The instrument was designed to receive microwaves, ultrashort radio waves, from a communications satellite orbiting the Earth. One day, the two men were trying to clear up some static that was interfering with their work when they detected a strange, faint, hissing sound. Determined to track down the source of this interference, the two physicists climbed up inside the huge horn of the antenna. There they found what they believed to be the cause of their problems—pigons were nesting in their antenna. But even after they removed the birds, the hissing continued. It went on night and day. The sound came from every direction in the sky —from all over the Cosmos.

Penzias and Wilson were mystified by this peculiar noise. They could not account for it until they heard about

---

KARL JANSKY (1905–1950). *Karl Jansky is shown here at age 23 shortly after coming to Bell Laboratories from the University of Wisconsin. Jansky set out to find the cause of static noise on transatlantic telephone communications. He built a large, rickety radio antenna and mounted it on car wheels so that he could push it around in a circle to search for the troublesome interference. In 1931, Jansky discovered the source of the static to be radio emissions from outer space. In his honor, radio astronomers around the world today measure the intensity of radio signals from space in units called Janskys. (Bell Laboratories)*

the work of Robert Dicke. Dicke, a physicist at Princeton University, not far from the Bell Laboratories, was a supporter of the Big Bang theory and had also predicted that radiation from the fireball of creation should still exist in the Cosmos. He was, in fact, about to begin efforts to search for it himself when he realized that Penzias and Wilson had accidentally discovered it. The static coming from the Cosmos was due to radiation.

So, nearly 20 years after Gamow's prediction, evidence of the Big Bang, the afterglow of that superviolent event, was found. Astronomers quickly followed up this discovery and confirmed that the radiation was exactly the kind—the correct wavelength and temperature—that would remain some 15 billion years after a cosmic explosion. In 1978, Penzias and Wilson shared a Nobel Prize for this important contribution. Proof of the fireball's lingering afterglow clinched the debate over the Big Bang. Like it or not, astronomers now had to accept the Big Bang theory.

PHYSICISTS ARNO PENZIAS (1933–) AND ROBERT WILSON (1936–). In this photograph, Penzias (right) and Wilson of Bell Laboratories stand before the big horn of the radio antenna in Holmdel, New Jersey. The instrument, designed to receive signals from communications satellites, was hampered by strange noises coming from every part of the sky. This radiation proved to be important evidence supporting the Big Bang theory. (Bell Laboratories)

4

Further Evidence

T HE fireball radiation was not the only evidence to convince scientists that the Big Bang had actually occurred. For if the Big Bang theory was true, then the great explosion that created the Cosmos would have set into motion a process of change that is still continuing. As a result, the Cosmos should be very different now than it was some 15 billion years ago.

The oldest, most distant galaxies—the first ones to form after the explosive birth of the Cosmos—should be flying out into space at a faster rate than the more recently formed galaxies closer to us. This is so because gravity, the force that causes large masses to attract smaller ones, would have been nil for a time after the Big Bang, when only radiant energy existed. Only when matter had begun to form did gravity develop. When the first galaxies arose, therefore, gravitational force was still relatively weak. It had little effect upon the movements of the earliest galaxies. But as newer galaxies arose, cooled energy, now in the form of particles of gas and dust, already existed. This matter exerted a gravitational force pulling and tugging at the newer galaxies, slowing them down.

If it could be shown that those distant, older galaxies were indeed moving apart at a faster rate than the newer galaxies closer to us, then this would prove that real changes have taken place in the history of the Cosmos. This evidence would further confirm and strengthen the Big Bang theory.

But how can we learn if the Cosmos has been changing in the way that is predicted by the Big Bang theory? How can we find out what the Cosmos was like billions of years ago? We can look into the past with the giant telescope. For when we peer through a huge telescope at a distant star we are not seeing that star as it is today; we are seeing it rather as it looked in the past. This is so because it takes time for light to travel across the vast distances of space, from that faraway star to us here on Earth.

Therefore, when we observe the distant galaxies through a telescope we see them as they were when the light from them first started on its long journey to us. Since the speed of light is constant, when we look at a galaxy, say 100 million light-years away, we are looking back 100 million years into the past. In other words, we are seeing a galaxy as it appeared back when the dinosaurs lived on Earth. When we look through a giant telescope at a distant star we are like fossil hunters seeking evidence of the ancient past. We are using the telescope in a marvelous new way—as a time machine.

In 1959, Allan Sandage, an American astronomer, set out to investigate the whole question of change in the Cosmos by means of the giant telescope. Sandage hoped to learn whether or not the distant, older galaxies were indeed moving away at a faster speed than the nearer ones. If this was so, they would show a greater red shift. The degree of red shift is proportional to the speed of the galaxy. Sandage had the advantage of using the largest optical telescope in the world at that time, the huge telescope at the mile-high Palomar Mountain Observatory in southern California. This colossal instrument, which Hubble had helped to design, was completed in 1948. It weighs 500 tons and stands almost as high as an eight-story building. Its great reflecting mirror measures 200 inches across.

Night after night, Sandage took an elevator up to a position just under the dome of the observatory. From it he would crawl into the observer's cage located in a long tube at the top of the telescope. In this chamber he operated the controls that moved the instrument and brought the galaxies into focus so he could photograph them. Above him an open slit in the metal dome let starlight pass through into the telescope to the huge mirror below. The mirror reflected and focused the light up into Sandage's cage. He could not, however, photograph the most distant galaxies because their light was too dim. The

THE 200-INCH TELESCOPE AT MOUNT PALOMAR. *This colossal optical telescope weighs 500 tons and is almost as tall as an eight-story building. Its great, reflecting mirror is nearly 17 feet in diameter. Completed in 1948, the Mount Palomar telescope has served astronomers Hubble, Humason, Sandage, and others. The operator sits high above the barrel-shaped network of girders in a tubular chamber called the observer's cage. (Palomar Observatory Photograph)*

best he could do, and it was an extraordinary feat, was to record the light from galaxies 8 billion light-years away.

For 15 years Sandage continued his efforts. In 1965, when Penzias and Wilson announced their proof of the fireball radiation, the afterglow of the Big Bang, Sandage was still hard at work, locating galaxies and recording their distances and speeds.

By this time, engineers had developed an ingenious means of intensifying the very faint light from distant galaxies. This light-intensifying device, coupled to a computer, enabled the 200-inch telescope to perform as if it had a 2000-inch reflecting mirror. It extended the range of the instrument enormously. The device also permitted the operator to observe the sharp images of the heavens on a televisionlike screen in the comfort of a heated room. No longer did Sandage need to spend long, chilly nights huddled in the tiny viewing cage near the top of the giant telescope in order to photograph distant galaxies.

Finally, in 1974, Sandage reported his results. All told, he made careful and exact records of the red shifts of 42 galaxies. Although he was unable to observe the earliest galaxies at the very edge of the Cosmos, and had to settle for those that arose some 10 billion years after creation, Sandage confirmed that these older galaxies are moving

---

EDWIN HUBBLE IN THE OBSERVER'S CAGE OF THE 200-INCH TELESCOPE. The astronomer using the giant Mount Palomar telescope crouches in the observer's cage. Light from the night sky enters the telescope through a slit in the domed roof of the observatory. It is focused on the observer's cage by the instrument's great 200-inch reflecting mirror 55 feet below. (Palomar Observatory Photograph)

outward at a faster rate than the nearer ones. The Cosmos is continuing to expand but at a slower rate now than in the past. It is changing, just as predicted in the Big Bang theory.

This evidence should have settled the controversy about the Big Bang. Proof of the fireball radiation's existence, plus Sandage's findings that showed that the rate of expansion of the Cosmos has changed, add up to confirmation that there was a beginning and that the process of change is continuing. The cosmic countdown is in progress. No matter how appealing it is to believe that the Cosmos is eternal, unchanging, and will always be here, all the evidence points the other way. The Cosmos will not go on forever.

Still, it is not easy to accept that the Cosmos is impermanent, that it will fade into the shadows and vanish. Allan Sandage himself was unhappy with this outcome. "It is such a strange conclusion . . . it cannot really be true," he said. So, despite the evidence, some astronomers were still not at ease with the doomsday prediction of the Big Bang theory. They reluctantly accepted that the Cosmos had a sudden, violent beginning and conceded that the Cosmos had been changing. But, they argued, this process of birth and change might be part of a cycle that happens again and again.

According to this new concept, the Cosmos will expand just so far, before gravity finally prevails and brings the expansion to a halt. Then, the galaxies will be pulled back toward one another until the Cosmos collapses. A new cosmic egg will form to be followed again by another Big Bang and the creation of a new Cosmos. Those who hold this view see this process as a cycle repeated endlessly over and over again. In this scheme, the Cosmos never really comes to an end. It merely switches, or oscillates, back and forth between phases of contraction and expansion that go on forever. This concept is called the Oscillating Universe theory.

The Oscillating Universe theory is attractive. Albert Einstein, among others, had hopes that it might be proven correct. The difficulty with the Oscillating Universe theory, however, is that it depends on gravity to stop the expansion of the Cosmos and bring about its collapse. But astronomers have estimated that the total mass, the sum of all the stars, galaxies, and cosmic dust in space, and all the energy in the Cosmos, is not sufficient to halt the momentum of the outward-bound galaxies.

The known facts suggest that the Cosmos will continue to expand until the end of time, when the last star dies and the Cosmos ends. In spite of this, some determined astronomers still search for matter, unaccounted for so far, that might make the force of gravity strong enough to prove the Oscillating Universe theory correct and restore a sense of eternity to the Cosmos.

5

Invisible Signals

A STRONOMERS scan the heavens, searching for new truths about the Cosmos. These efforts to explore the Cosmos and learn its secrets are a remarkable tribute to human ingenuity.

Cosmology, the study of the Cosmos, was hardly a science at all when it began centuries ago. It was an interest based on observation, speculation, and imagination. And since observation was restricted by the limits of human eyesight, cosmology was mostly guesswork. Still, the history of astronomy is a splendid record of intelligent, sometimes brilliant, efforts to grasp at truths beyond knowing and worlds beyond view.

Consider the handicaps of the astronomer. Unlike the biologist, the astronomer cannot dissect the subject of study to learn its structure. Nor can the astronomer readily measure or weigh it in a physics laboratory, or test its properties in a test tube, like a chemist. Unaided, the astronomer can see only a tiny fraction, perhaps a few thousand, of the 200 billion stars in our Milky Way Galaxy. How then, except for the invention of the telescope, could this ancient interest in the heavens ever have developed into the most exciting science of our time? The answer is that it could not.

The telescope was probably invented by a Dutch lens grinder, but its development as an astronomical instrument is attributed to the Italian Galileo Galilei. In 1610, when Galileo first trained his homemade telescope on the heavens, modern cosmology was born.

But the optical telescope, with all its gadgetry and perfection, still has important limitations. Its effectiveness depends upon favorable weather conditions. It requires clear, dry, cloudless skies and can be used only at night. Turbulence in the upper atmosphere—the disturbance that causes stars to appear to twinkle—can also interfere with the sharpness of the telescopic image, and so can air pollution. Light from nearby cities can produce a bothersome haze.

Technology, however, has not stood still. It has produced alternatives to the optical telescope. New instruments detect forms of energy, other than light, radiating through space. For visible light is only one of several kinds of energy in the Cosmos. These different types of energy travel through the vast reaches of space in the form of waves, like ripples in water. Each type has a specific wavelength and frequency, or number of ripples per second. The longer waves have fewer ripples per second than the shorter waves. Scientists classify the various types of energy by arranging them in order, according to their wavelengths.

This arrangement is called the electromagnetic spectrum.

At one end of the electromagnetic spectrum are the longest waves, the radio waves. Then come the microwaves, the infrared heat waves, the narrow band of visible light, ultraviolet, x-rays and finally, the shortest waves, the gamma rays. A radio wave may be miles long, but a gamma ray, at the other end of the electromagnetic spectrum, may only be a trillionth of an inch in length. Despite their differences, however, all these forms of energy travel through space at the same speed, about 186,000 miles per second.

But not all forms of cosmic energy reach the Earth. Fortunately for us, only the shorter radio waves, microwaves, and visible light waves penetrate the relatively dense, gaseous atmosphere that surrounds our planet. The other radiations, which are harmful to us, are largely screened out before reaching the surface of the Earth. Still, some ultraviolet rays do get through and affect us, as you know, if you've ever been sunburned on a clear summer afternoon.

Until 1931, just about all the knowledge we had about the Cosmos was gained from observations of visible light, starlight. Then Karl Jansky accidentally discovered that radio waves from space reach the Earth. At first, Jansky's discovery had little effect—no one seemed to realize its

THE ELECTROMAGNETIC SPECTRUM. Most of the information we have about the Cosmos has come to us in the form of light from distant stars and galaxies. But light is only one of several kinds of electric and magnetic forces that travel through space from far-off stars in the form of waves. Each of these different types of energy has a specific range of wavelengths. In the electromagnetic spectrum shown, these various kinds of radiation are arranged according to their wavelength.

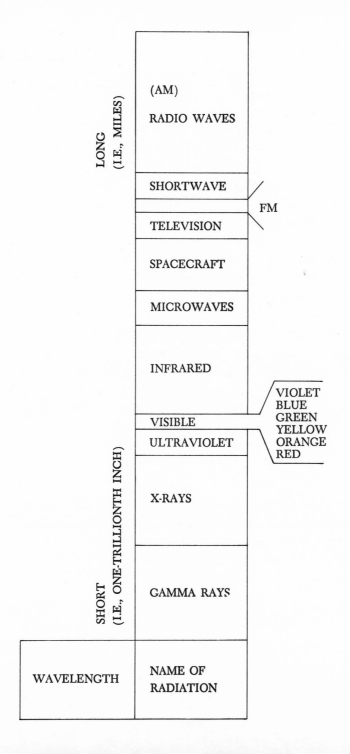

importance to astronomy. Jansky, after all, was an electrical engineer and not a professional astronomer. But then Grote Reber, another young electrical engineer, learned of Jansky's finding and became interested. Reber lived in Wheaton, Illinois, a suburb of Chicago. He was an amateur radio operator, a "ham." Since his teens, his hobby had been radio, and he communicated frequently with other amateur broadcasters.

Reber was fascinated by Jansky's discovery and set about building an antenna to see if he too could detect these mysterious radio waves. So he went out and built a large dish-shaped antenna, 31 feet across, in his backyard and aimed it skyward. In 1938, after some difficulties, Reber too heard the hum of the Cosmos. He confirmed Jansky's findings. Reber had designed and constructed the first dish-shaped radio antenna.

Still, astronomers were little interested in these invisible signals from space. When the Second World War broke out in Europe (1939–1945), however, cosmic radio waves came to attention again and again as static that interfered with the reception of radar signals. The British had built great radar towers along their Channel coastline to detect enemy aircraft. But static interference often hampered the effectiveness of the radar. At first, the British believed that the Germans were deliberately broadcasting signals to jam their detection equipment. But the static turned out to be bursts of radio waves coming from the sun.

Careful not to reveal their discovery lest Germany learn more about radar, the British kept it a secret until the war ended. Then scientists and engineers who had

THE 250-FOOT RADIO TELESCOPE AT JODRELL BANK, ENGLAND. *The huge, dish-shaped reflector of the radio telescope functions much like the reflecting mirror of an optical telescope. But it gathers radio waves instead of light waves and focuses them on an antenna. The antenna in this photograph is the small tower protruding from the center of the reflector. The radio telescope at Jodrell Bank can be turned to scan any part of the sky. (University of Manchester)*

worked on the British radar system constructed new antennas designed specifically to receive radio signals from space. Radio astronomy had been born. By 1948, when the huge 200-inch optical telescope at Palomar Mountain became operational, enormous radio telescopes were also under construction. Sadly, however, Karl Jansky died in 1950, at the age of 44, from a kidney disease, just when his discovery was to herald a new era in astronomy.

Soon radio astronomers were constructing enormous antennas all around the world: in Australia, West Germany, England, Sweden, Canada, Russia, and the United States. The antennas differed in design; some were round, dish-shaped, sheets of metal, while others were huge, bowl-shaped grids. The largest single radio antenna in the world was built in a valley in western Puerto Rico. It is 1000 feet wide, or the size of more than three football fields laid end to end.

Some radio telescopes consist of separate, individual antennas spread over many miles and linked together electronically. Such a multiple antenna has the power to act as if it were a single instrument many miles wide. Its individual units are all operated by remote control and

---

THE 1000-FOOT RADIO TELESCOPE AT ARECIBO, PUERTO RICO. Near the western end of the island of Puerto Rico, engineers took advantage of a natural bowl in the hills to construct a gigantic dish of wire mesh. Though the reflector is fixed and immobile, it is able to receive radio waves from many directions because of the Earth's rotation. Additional flexibility in scanning is obtained by moving the antenna-receiver that is suspended from cables above the face of the dish. (NAIC/ Cornell University/NSF)

directed with computer assistance from a headquarters building. Such an arrangement is the Very Large Array, or VLA, built on a plain in New Mexico. The VLA is the world's largest radio telescope. It consists of 27 antennas spread over an area of about 21 miles. The antennas are set on a Y-shaped railroad track so that the spaces between them can be adjusted.

Why must radio telescopes be so large? Optical telescopes rely on visible light which has relatively short wavelengths compared to radio waves. To gather the much longer radio waves requires very large antennas. Besides the antenna, all radio telescopes have an electronic receiver. Radio waves are extremely faint. The electronic receiver strengthens, or amplifies, these weak, invisible signals. Radio waves can be made audible. They sound like soft hissing or humming. The signals can be inscribed on graph paper and recorded on magnetic tape.

Anything that interferes with the reception of the faint radio waves from space affects the radio telescope. Ordinary weather conditions, like rain, fog, or snow, have little effect, but lightning or electrical storms do hamper operations. So does human-made electrical "noise" such as that produced by airport radar towers, high-voltage hospital equipment, and even faulty automobile ignition systems. For this reason, radio observatories are purposely built in remote places away from cities and population centers.

Where do these mysterious radio signals from space originate? Radio astronomers are discovering thousands of origins. They come from extragalactic sources, from objects beyond our Milky Way Galaxy. These great, distant, mysterious structures that are strong sources of radio waves are called *radio galaxies*—they are the largest known bodies in the Cosmos. Radio waves signal changes and violent events in the Cosmos. They are generated in explosions on the surface of the sun and in bursting stars. They also come from planets and from swirling clouds of gas and dust between the stars. These clouds of dust that may block light and make optical telescopes blind are no barrier to radio waves. The radio telescope can "see" through them easily.

Astronomers have been able to learn the shape of our Galaxy with the aid of radio telescopes. This is a remarkable achievement when you consider that it is not yet possible to venture beyond our Galaxy to view it from the outside. This is roughly like trying to learn the shape of your home or apartment house, never having been outside one room within it.

The great, sensitive antennas of radio telescopes detect radio waves emitted by hydrogen gas floating in interstellar space, the space between the stars. By recording these radio emissions and charting their distribution, radio astronomers learned that this "radio glow" is not evenly spread throughout the Galaxy. Rather, they found that the interstellar gas is concentrated in lanes that are separated by wide, relatively empty zones. So a radio map of the interstellar gas, the "radio glow," gives us a picture of the Milky Way, an image of a vast spiral with a number of arms.

6

Red Giants
to White Dwarfs

I n all the Cosmos nothing seems more eternal than the stars. Generations upon generations of people have observed the stars, and they appear unchanged. But the stars did not always exist, nor will they last forever. Stars are born, mature, and die like living things. They age slowly, however, lasting millions, and, in some cases, billions of years. There are stars in the Cosmos that date back to shortly after the Big Bang, and there are new stars forming out in space at this very moment.

How is it possible, given the relatively brief life of people, for us to understand the life cycle of a star? The whole record of the human race is less than a blink of an eye in the life of a star. We cannot, therefore, observe the course of any one star from birth to death to learn its history—nor is that necessary. For when we observe a typical human community, what do we see? We see people at all stages of life: babies, children, teenagers, young adults, the middle-aged, and elderly folks. And so we understand the human life cycle. The galaxies in the heavens are similar communities. They are neighborhoods teeming with stars of all ages. By studying them through telescopes we come to recognize their young and their old, and the life cycle of the stars is revealed to us.

Stars arise from the same seeds that generate all matter in the Cosmos. As the fading fireball of creation dimmed some 15 billion years ago, a great, primitive cloud of gas and dust filled the Cosmos. This gas and dust were made up of tiny bits of matter called protons and electrons. These particles came together to form atoms of the simple gas hydrogen. Each hydrogen atom consists of one proton that accounts for almost all its mass, and an infinitely smaller electron that circles around it. Hydrogen is the most abundant element in the Cosmos, making up 90 percent of all matter. Helium, formed from two atoms of hydrogen, is the second most common element found in the Cosmos.

Thin clouds of swirling hydrogen gas fill space. Now and again in the movements of these clouds atoms are temporarily swept together. Normally, their chance motions separate them quickly once more. But each atom exerts a weak gravitational pull on its neighbor. When, by accident, large numbers of atoms happen to come close, their combined gravity may be strong enough to keep them from moving apart again. When this occurs a separate, permanent pocket of gas forms.

Over time the force of gravity continues to pull the atoms of the pocket of gas closer to one another. As the atoms are drawn together, they travel inward faster and their energy increases. This produces heat, warming the center of the pocket. More and more atoms are pulled

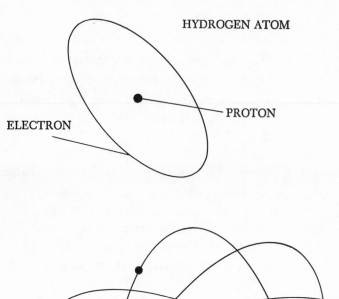

HYDROGEN ATOM

PROTON

ELECTRON

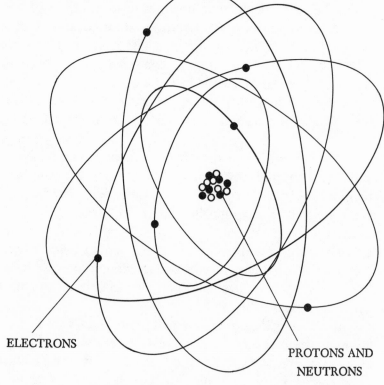

ELECTRONS

PROTONS AND
NEUTRONS

CARBON ATOM

inward, causing the gas pocket to contract and grow still hotter. After some millions of years of contraction, the pocket is reduced to a fiercely hot ball of compressed gas. It has become an embryonic star.

The ball of gas continues to shrink under the influence of gravity. Its core grows hotter still. After some millions of years of compression, the temperature at the center of the ball finally reaches 20 million degrees Fahrenheit. At this extreme heat, about 1000 times the temperature of the hottest steel furnace on Earth, the hydrogen bursts into flame. The core of the compressed ball becomes a fiery inferno in which excited protons race about, smashing into one another. These violent collisions cause the protons to stick together, or fuse, releasing vast amounts of nuclear energy. The fusion of protons creates the awesome power of exploding hydrogen bombs. Such are the first stirrings of a newborn star.

THE ATOM. All matter is made up of three basic building blocks: a light particle called the electron and two relatively heavy particles called the neutron and the proton. The neutrons and protons are bound very tightly into a compact mass called the nucleus.

Electrons possess negative charges and are attracted to the nucleus by the positive charge of the protons. The electrons circle in orbits around the nucleus under the influence of this attraction. Together the nucleus and orbiting electrons form the atom. The atoms diagramed here are those of hydrogen and carbon. The hydrogen atom consists of one proton and one orbiting electron. The carbon atom has six protons and six neutrons in the nucleus. Six electrons circle the nucleus. The diameter of a typical atom is one hundred-millionth of an inch.

The infant star is a fiery ball a million miles in diameter—about the size of our sun and other medium-size
stars. The vast energy produced at its center by the burning hydrogen passes outward to the surface of the star and
radiates away into space in the form of light and heat. The
passage of this nuclear energy from the core outward op-
poses the force of gravity pushing inward, keeping the
star from contracting further. Ninety-nine percent of the
star's life is spent in this hydrogen-burning, or fusing,
stage.

The fusion reactions at the center of the star produce
a new element, helium. While the star burns hydrogen
and produces helium, its appearance changes little. It remains yellowish in color. But as helium accumulates at
the center of the star and the supply of hydrogen dwindles,
signs of aging become apparent. The star's outer layers
begin to swell. Gradually they turn from yellow to orange
and then, more quickly, to red. At this stage the aging star
bloats into a huge, red ball 100 times larger than it was at
birth.

Our sun was born about 4.6 billion years ago and is
now halfway through its hydrogen-burning stage; it is in
midlife. In another 5 billion years or so, the sun is expected
to reach the swollen, red stage. By then it will have bal-
looned into such a great ball of gas that it will swallow up
the planets Mercury and Venus, and reach out almost to
the Earth.

Viewed from the Earth, the red sun will fill most of
the sky. But we will not be able to witness this spectacle
directly because the sun's rays will have increased the
temperature at the Earth's surface to some 4000 degrees

Fahrenheit. Life will have vanished on Earth, and the planet's surface will have been reduced to molten rock. The oceans will be gone, evaporated away. It is to be hoped, however, that we will be safely resettled elsewhere in the Galaxy by then.

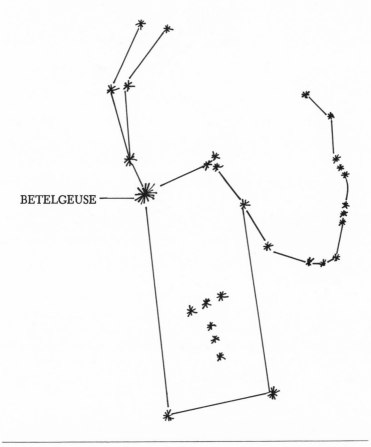

BETELGEUSE

ORION, THE HUNTER. *This drawing shows how stars in the constellation Orion form the shape of a hunter. The red giant Betelgeuse is marked.*

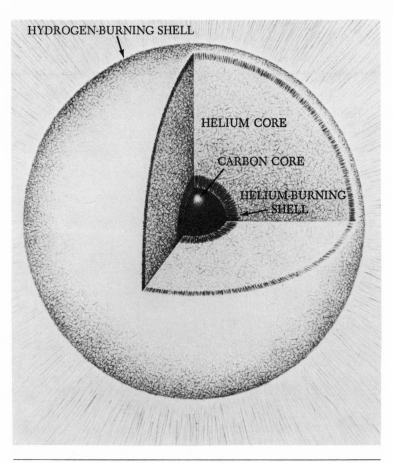

THE STRUCTURE OF A STAR LATE IN LIFE. At the beginning of a star's life it burns hydrogen to make helium. In the later stages of its existence, helium burns to make carbon. This illustration shows the interior of a star during the later period in which carbon is accumulating at its center. The carbon is surrounded by burning helium; the helium is surrounded, in turn, by a layer of hydrogen that has been burning since the star was born. (Adapted from Astronomy: Fundamentals and Frontiers by Robert Jastrow and Malcolm Thompson, published by John Wiley and Sons)

Astronomers call these aging, swollen stars *red giants*. In the constellation Orion, the Hunter, the bright star Betelgeuse (pronounced Beet-il-jooz) is a red giant. You can see its red tint easily on the right shoulder of the hunter on a clear, winter night, looking southward in the Northern Hemisphere.

Red giants live out their final years, slowly burning up their dwindling hydrogen. When they exhaust their fuel, the nuclear energy that has kept the star from collapsing gradually gives way to gravity. The star begins to shrink once more. As it contracts, however, it produces more heat. This heat stirs the helium atoms that have collected at the center of the star. They become excited, and violent collisions now take place between them. A process begins like that which ignited the hydrogen earlier.

But helium fusion requires a temperature of 200 million degrees Fahrenheit, 10 times that of hydrogen, to burn. When this critical temperature is reached, the core of the star reignites. It becomes a nuclear inferno once more. Helium nuclei fuse to form the element carbon. The nuclear energy released now halts the further collapse of the star. And the star gains new life burning helium.

All stars age in a similar way, first burning hydrogen, then helium. Beyond this point, however, the fate of stars depends on their size and mass. Small stars are crushed by gravity and wither away. Large stars disintegrate in great explosions.

For our sun and other average-size stars, the helium-burning stage lasts about 100 million years. By then, the helium supply at the center of the star is used up, and the core fills with carbon. A further renewal of the star's life would require atoms of carbon to fuse. But for carbon fusion to take place, the temperature at the center of the contracting, self-heating star must reach 1 billion degrees Fahrenheit, five times the heat it took to fuse helium atoms.

Temperatures at the core of shrinking red giants cannot reach the critical heat required to fuse carbon atoms. So the force of gravity prevails once more, and the dying star collapses under its own weight. A medium-size star, like our sun, contracts to one millionth of its original size, down to a sphere similar in diameter to that of the Earth. Gravity compresses matter in the star so tightly that a teaspoonful of its core would weigh 10 tons.

Although the interior of a collapsing star never gets hot enough to fuse carbon, the self-heating process continues as the star contracts. The surface temperature of the dying star increases, and the scarlet glow of the red giant gives way to white-hot heat. Astronomers call these shriveled, densely compacted bodies *white dwarfs*. Without sufficient heat to fuse carbon atoms, however, white dwarfs lack new sources of energy. Eventually they cool down, and, as they do, their color turns gradually from white to yellow, to dusky red, and then to black. At their death such stars are reduced to cold, dark lumps called *black dwarfs*, lifeless bits of debris in the Cosmos.

7

Stardust

A LL stars do not share the same fate. The larger, more brilliant ones burn their supplies of hydrogen and helium more rapidly than average-size stars like our sun. As a result they live shorter lives, millions of years instead of billions. Nor do the larger stars simply continue to collapse, shrivel up, and become white dwarfs.

Because of its great mass, the collapse of a large star generates far greater heat than does the final contractions of smaller stars. Large stars do not become white dwarfs. Rather, a dying, large star produces heat that rises to that critical level at which carbon atoms fuse, 1 billion degrees Fahrenheit. These carbon fusion reactions produce new, heavier elements. And even when the star exhausts its carbon fuel, the cycle of contraction, heating, and nuclear reaction continues. The core of the star manufactures more elements, 26 of them. Ranked in order of their increasing atomic number, the number of protons in the nucleus, these elements include: nitrogen, number 7; oxygen, 8; fluorine, 9; and so on, up to and including iron, number 26.

Iron, however, resists fusion. It absorbs the energies generated at the center of a great star. So, when iron builds up at its core, a large star is doomed. The nuclear furnace cools, and the energy that had opposed gravity gives way, leading to the star's final collapse. Gravity sucks matter into the dense center of the star, and material flies toward the core, traveling at a speed of more than a million miles an hour. Matter accumulates and becomes tightly packed. It pinches the core of the star, creating great pressure. This pressure increases until finally it halts the further collapse of the star.

The death of a great star is a spectacular event. The squeezed core of the star can no longer contain the enormous pressure that has built up within it. So, suddenly, like some great cosmic jack-in-the-box, this pressure pushing outward causes the unstable star to erupt. The star explodes with unimaginable power. After a life that has spanned millions of years, the death of a large star, from its final collapse to its violent disintegration, takes only a few minutes.

When the star blows up, it spews into space all the elements it has manufactured throughout its lifetime. But the final explosion itself generates trillions of degrees of heat. This heat triggers, in turn, countless nuclear reactions, and the energy released by them causes still others. In a few seconds, these chain reactions produce all the natural elements found in the Cosmos.

The relative abundance of the different elements depends on when in the life of the star they were made. Carbon, oxygen, and the other elements up to iron, atomic number 26, accumulate for millions of years before the death of great stars—these elements are abundant in the Cosmos. Silver, 47; platinum, 78; gold, 79; and the heavier elements up to uranium, atomic number 92, are created, however, only in the final few catastrophic seconds of the star's life; that is why they are scarce.

The elements scattered into space by the explosion of a great star mix with fresh hydrogen. Over time new stars form from clouds of hydrogen that have been enriched by this debris. These new stars, in their lifetimes, manufacture more elements. This cycle is ongoing. Each great star that dies leaves behind a legacy in the form of material to be incorporated in new stars. Our sun is a star created from such stardust. The planets in our solar system are also made up of this material, so is our Earth. Even you and I are composed of stardust—elements from exploded great stars that have been recycled in the Cosmos over billions of years.

Astronomers call exploding large stars *supernovas.* The ancients, who observed and described such events, believed that they were witnessing the appearance of a new star, or "nova." Today we recognize that what the ancients actually saw with their unaided eyes were not new stars being born but large, old stars dying. The term "super" refers to something great, or large, and so the name supernova. Supernovas shine with a brilliance many billions of times brighter than our sun. Therefore, were a supernova to appear in a neighboring part of our Galaxy, it would be clearly visible even in daytime.

At least three appearances of supernovas have been recorded in history. The most recent, in 1604, was observed and described by the German astronomer, Johannes

Kepler, and by the Italian, Galileo Galilei. Before that the Danish astronomer, Tycho Brahe, witnessed one in 1572. Still earlier, Chinese astronomers noted the occurrence of a similar event in the constellation Taurus, the Bull, in the summer of the year A.D. 1054. Today, more than 900 years later, evidence of the supernova described by the Chinese is still visible. At the location in space where the explosion took place there is a bright, glowing cloud called the Crab Nebula. The Crab Nebula is a great pocket of luminous gas created when the supernova blew up, spraying its hot contents into nearby space. The Crab is expanding outward at the speed of 700 miles per second. It will eventually disperse, spreading materials from which new stars will arise.

But what happens to the core when a star explodes in a supernova? Does the core disintegrate and also blow away with the gaseous outer layers of the dying star? Astrophysicists (physicists who study the Cosmos) wondered if some remnant of the supernova's core would remain

---

THE CRAB NEBULA. *The Crab Nebula, the luminous cloud in this photograph, is the remains of a great star that exploded more than 900 years ago, in* A.D. *1054. The deaths of great stars are spectacular events. The exploding star, called a supernova, shines with a brightness billions of times greater than the sun. If the supernova is nearby, in our Galaxy, it appears suddenly as a new star so brilliant that it is clearly visible in the daytime.*

The beautiful, lacy Crab Nebula contains all the elements manufactured during the life of the star. It is expanding rapidly outward into space spreading debris that will eventually be recycled, giving rise to new stars and planets. All the elements that form our Earth and all living things—including people— contain the stardust of earlier exploded stars. (Lick Observatory)

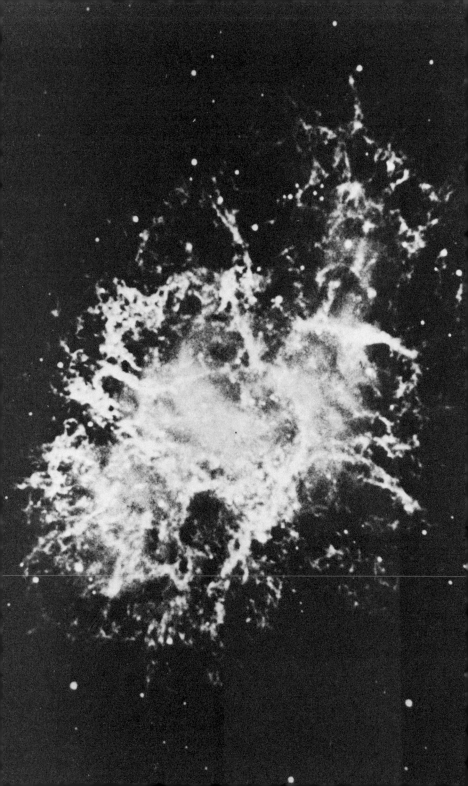

intact. As far back as the mid-1930s, some astrophysicists became convinced that the extremely compacted core of the supernova would survive the star's destruction.

Electrons, circling the nucleus of an atom, have small, negative electrical charges. Protons, within the nucleus, carry positive charges. So some scientists predicted that electrons, driven into the nuclei of the atoms by the force of an exploding supernova, would join protons there to form electrically neutral particles called *neutrons*. These scientists thought that the core of an exploded supernova would consist of a tightly packed ball of neutrons. This remnant would be extremely dense matter.

A massive star, many times larger than the sun, might leave behind its original core condensed to a relatively tiny sphere, perhaps only 20 miles in diameter, much smaller by far than a white dwarf. To get some rough idea of the relative densities of these bodies, imagine that we have at our disposal a cosmic scale. A teaspoonful of the core from a white dwarf would weigh 10 tons on our scale. A similar amount of matter from the core of an exploded supernova would weigh 10 billion tons. This extraordinary cosmic corpse, this ball of neutrons, was given the name *neutron star*.

Astronomers debated the existence of neutron stars for 30 years. They searched the skies for them. They hunted for one in the Crab Nebula where the shrunken hulk of the supernova explosion of A.D. 1054 might be expected. But the elusive neutron star was not to be found.

Interest in neutron stars began to wane. In August, 1967, however, an extraordinary find was to change that. Like the discoveries of Slipher, Jansky, and Penzias and Wilson earlier, this find, too, was accidental. In collecting data for another purpose, a research astronomer came upon something totally unexpected. At Cambridge University in England, a team of astronomers was studying fluctuations in radio waves coming from galaxies far off in distant space. They were searching the Cosmos with a new, highly sensitive radio telescope.

A graduate student member of the team, Jocelyn Bell, was assigned the task of examining the rolls of paper on which the telescope's input was inscribed in wavy lines forming peaks and valleys. One day the alert young woman noticed a curious difference in the signals being recorded. They were not at all like those with which she was already familiar. The strange signals were short, rapid bursts of radio waves, occurring at remarkably precise intervals. They were so regular that they might for all the world have been the rapid dots and dashes of some cosmic Morse code. Each recorded flicker lasted no more than one hundredth of a second.

The newly discovered radio waves puzzled the Cambridge team. No known body in the Cosmos emitted similar signals. What mysterious source was sending them? Was it possible that they were communications intentionally beamed to Earth from some other intelligent civilization out in the Cosmos? At first, considering this a serious possibility, the Cambridge group decided to keep their finding secret. They decided it was better to investigate further before going public with it.

Among the members of the Cambridge group, the still-unidentified source of the signals was known as LGM, or Little Green Men. It soon became apparent, however, that the strange radio pulses had a natural origin. They were spread over far too wide a band of frequencies to have been intentionally produced for communication purposes. Disappointed, perhaps, the astronomers reported their discovery, and LGM was soon replaced by the more scientific-sounding term, *pulsar*. By 1968, researchers had discovered pulsars, bodies that emit regular radio pulses, in other parts of the Cosmos also.

But what were these pulsars that flashed on and off, like radio beacons in space? After examining various possibilities, astronomers agreed that a pulsar must be an extremely small, rotating object to be capable of producing such sharp, rapid, and regular pulses. By calculation, they estimated that a pulsar would turn out to be no larger than a mass 20 miles in diameter, by cosmic standards a microscopic body. Previously, white dwarfs were generally regarded as the smallest, most compact bodies in space. Now, along came these curious pulsars, a thousand times smaller.

The world of astronomy buzzed with interest in pulsars, and the stage was set for a dramatic surprise. It came in 1968 with the discovery of a pulsar located at the heart of the Crab Nebula, the very site where, 34 years earlier, astrophysicists had forecast that a neutron star should exist. Now, quickly, the known facts and theory were matched and assembled like pieces of a complex puzzle. The pulsar discovered in the Crab Nebula was the size expected of the long-sought neutron star, the compacted core of the supernova that exploded back in A.D. 1054. A *pulsar and a neutron star were one and the same thing*. A remarkable breakthrough had been achieved. The basis for much theoretical research in astronomy was strengthened.

Attention turned next to the question of just how a pulsar, or neutron star, emits radio signals. Current theory has it that violent storms take place on certain surfaces of the superdense star, like those that occur on the face of the sun and other stars. The bursts of signals are thought to result from showers of radioactive particles swept into space by these storms. When the Earth happens to cross the path of this radiation, our radio telescopes detect it, and we become aware of a pulsar.

But what accounts for the sharpness and the regularity of these radio flashes? Being extremely small and rotating very rapidly, it is believed that the pulsar emits bursts of radioactivity that gather into narrow bands like the beams of light from a revolving beacon. When the Earth lies in the path of this rotating beam, it receives a sharp flash of radiation with each quick turning of the tiny pulsar.

Will all great stars become pulsars, or neutron stars? Many astronomers believe so, and for them, the story of the birth and death of stars ends here. But other authorities are convinced that the fate of some truly massive stars has still another final chapter, an ending even more mysterious and bizarre.

8

Black Holes in Space

THE secrets of star formation, red giants, white dwarfs, supernovas, and neutron stars (pulsars) were hard won, but questions about the death of stars remain. Are neutron stars the final, most condensed, form of star stuff in the Cosmos? Could the most massive stars meet another, still stranger, fate? Some authorities think so. They suggest that the very largest of the great stars have a destiny so bizarre, in fact, that it strains the imagination.

What happens to these gigantic suns? Some astronomers think that when an exceedingly massive star dies its final collapse can compress the star's core down to a body even smaller than a neutron star. Nothing known to science approaches the density of such a mass. If the shrinking core is pinched down to a radius of only 2 miles, it would have an incredible gravitational force, billions of times greater than that of ordinary stars.

The crushing gravitational force would produce a strange effect. It would snuff out all light emanating from the core. Light produced by collisions between the jammed-together neutrons in the core could not escape; it would be trapped within. The last remnant of a massive, brilliant star would be plunged into everlasting darkness.

This peculiar phenomenon, the force of gravity trapping light, was predicted by Einstein's Special Theory of Relativity (1905). Among other things, this theory proposed that all matter and energy are equivalent. Einstein's equation $E=mc^2$, or energy (E) equals mass (matter) (m) multiplied by the speed of light (c) times itself, shows the relationship between energy and matter. Energy is matter that has been freed, or let loose, as in an atomic bomb. Matter, on the other hand, is energy that is bound up, or locked in.

Light is a form of energy and therefore can be said to possess a certain mass. A ray of light can be affected as though it were a stream of particles. It can be influenced by gravity. The force of gravity tugs at each ray of light that leaves a star, just as it pulls at a ball thrown up from the Earth, forcing it to fall back. But the force of gravity, spread out over the vast surface of even an average-sized star, like our sun, is not strong enough to keep light rays from escaping into space. The star is visible; the sun shines.

But what happens to the exceedingly dense relic of the great, massive star? According to the predictions of Einstein's theory and the laws of physics, as they are currently understood, the compacted, now darkened, core continues to shrink. It becomes what astonomers call a *black hole*. From a lightless body a few miles across, the contracting core is reduced to the size of a BB, a speck, something microscopic. Yet, this is not the end. The core, the great heart of a once colossal star, will become unimaginably tiny. But it will never disappear. For, according to what is known, neither energy nor mass can ever be destroyed.

Black holes, strange, infinitesimal remnants of giant stars, all but defy reason. Crushed to almost nothing, a black hole still has the mass of a star, perhaps a thousand trillion trillion tons of condensed matter packed into it. This gives a black hole an incredible gravitational force.

Black holes not only keep light trapped, but they attract and capture other objects close by. Each day, black holes may devour several nearby stars as well as other surrounding matter. Like some scary monsters of science fiction, the awesome appetite of the black holes keeps growing. As matter accumulates within them, the black holes' mass increases, their fields of gravity become stronger, and reach out farther into space. Unseen black holes may be hungrily eating up chunks of our world at this very moment. Will our Cosmos be devoured by them?

Are black holes a menace to future space exploration? Chances are that in the vastness of the Cosmos spaceships would travel safely well beyond the gravitational range of black holes. But what if a spacecraft ventured closer? What would happen if, by accident, an astronaut were to take a space walk and head unawares toward a black hole? If he/she approached head first, his/her head would pop off like a cannon ball to be followed by his/her torso and limbs. Feet first, the astronaut would be stretched like someone on a medieval torture rack for a few thousandths of a second while his/her limbs and trunk were pulled away. Either way he/she would be dismembered and sucked up in a wink.

Do black holes really exist? Not all astronomers agree

on this. But the same laws of physics that have enabled us to unlock so many secrets of the Cosmos, and calculations based on the life cycles of massive stars, suggest that black holes must exist. Is there any evidence of them?

Since space is so vast and black holes are so minute—and not visible—finding one would seem impossible. But clues to the existence of black holes have come from new technological developments in astronomy. The advances of the Space Age have opened new windows on the Cosmos. Now man-made satellites, carrying sensitive instruments, orbit the Earth and venture far out into space, providing us with important new insights.

Instruments placed in orbit above the Earth detect high-energy waves that are blocked by the planet's gaseous atmosphere. These forms of energy have very short wavelengths and are located at the other end of the electromagnetic spectrum from the long radio waves. The short, high-energy waves include ultraviolet, x-rays, and gamma rays. In 1970, the National Aeronautics and Space Administration (NASA) launched an unmanned satellite from a platform in the Indian Ocean near Kenya, East Africa. It was named *Uhuru* (the Swahili word for "freedom"), and it was designed to detect x-ray sources in space. One especially powerful x-ray source *Uhuru* located seemed to be a very bright star in the constellation Cygnus, the Swan.

In 1973, however, another satellite, *Copernicus*, revealed that the x-rays were not coming from the star at all but from an *invisible* object nearby. Astronomers think that this invisible object is a black hole, circling around the star, sucking up streams of gaseous particles being torn from the star's surface. They believe that these atoms, pulled in swift currents, bump and crash into one another before disappearing into the black hole. These atomic collisions are thought to generate temperatures of millions of degrees at the brim of the black hole, producing x-rays and gamma rays. No other explanation fits the circumstances as well.

A black hole may lie at the center of our own Milky Way Galaxy. The hub of the Galaxy, shrouded from view by clouds of dust and gas, was thought to be relatively calm, like the serene center of a gently swirling sea. Now, however, new techniques permit astronomers to "see" through the murk. They have found that, far from being serene, the center of the Galaxy is the site of great violent explosions and turbulence.

The great clouds of gas and dust at the core of the Galaxy are swirling around at speeds of millions of miles an hour. Could this whirlpool be caused by an extremely intense gravitational field, such as that surrounding a black hole? Other studies hint this could be the case. Astronomers report an exceedingly active source of radio waves at the core of the Galaxy. Infrared and gamma-ray detectors also point to the same site as a high-energy center.

The stars that make up a galaxy tend to cluster most densely about the galaxy's center. They are jammed closely together there in a relatively small area, and so they often collide violently. This condition favors the development of a black hole because, if a great, massive star burns out in this crowded hub, there is much debris for the black hole to feed upon. In fact, once it has formed, the black hole will become surrounded by a brilliant, luminous zone. This extremely bright halo is caused by the glowing heat of nearby stars, gaseous matter, and debris being forced together and drawn onto the surface of the black hole. At least, this is what some astronomers currently think explains another strange phenomenon, the *quasar*. Quasars, or quasi-stellar bodies, are the most mysterious, luminous objects in the Cosmos.

In 1960, Allan Sandage was trying to identify the source of certain powerful radio emissions when he made an unusual discovery. He was scanning the area from which the strong signals were coming with the 200-inch optical telescope at Mount Palomar when he located a tiny, bluish object different from any known body in space. Soon other optical astronomers reported that they too had identified similar starlike, or quasi-stellar, bodies.

Three years later, in 1963, a Dutch-born astronomer, Maarten Schmidt, discovered another curious feature of quasars that made them even more puzzling. He was studying quasar 3C 273 (the designation represents the quasar's classification, number 273 in the *Third Cambridge Catalogue of Radio Sources*) with the 200-inch optical telescope, and found, to his amazement, that it was far beyond the distance of any other object in the Cosmos. Its red shift placed it about 2 billion light-years away, and it was retreating at great speed. When the distance to the quasar 3C 48 discovered earlier by Sandage was checked, sure enough, it, too, showed a marked red shift. Sandage's quasar was even more distant, about 4 billion light-years from Earth.

More recently, quasars have been discovered 15 billion light-years away. Even in cosmic terms, that distance is very far, approaching the edge of the Cosmos. To be seen at such a distance, the light of a quasar must be of incredible intensity, far greater than that of a galaxy with millions of suns—it is roughly equivalent to the brightness of a hun-

3C 48

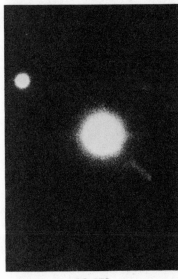

3C 273

QUASARS. *Quasars are among the most puzzling objects in the Cosmos. For a long time these faint bodies were regarded merely as dim stars in our Galaxy. In 1960, however, astronomers discovered something that set these starlike bodies apart from other stars. These dim objects emitted strong radio waves. Stars are not sources of radio signals.*

Then, in 1963, astronomer Maarten Schmidt checked the red shift of quasar 3C 273 and was surprised to find that this strange object was not in our Galaxy at all but was 2 billion light-years away. Quasar 3C 48, discovered in 1960 by astronomer Allan Sandage, turned out to be 4 billion light-years from Earth. Other quasars proved to be even more distant—up to 15 billion light-years away. Taking account of these vast distances reveals that the dim-appearing quasar actually shines with a brilliance hundreds of times brighter than a typical galaxy containing billions of stars. But a quasar is far smaller than a galaxy; it is not much larger than our solar system. (Palomar Observatory Photograph)

dred galaxies. Yet, amazingly, quasars are not much larger than our solar system—8 billion miles across—very small on a cosmic scale.

Using the telescope as a time machine, we discover that the distance of the farthest quasars takes us back in time to the first few billion years after creation, long before the sun and the Earth were born. Astronomers believe that these distant quasars were associated with massive black holes that formed at the centers of the early galaxies. The quasars we see today no longer exist. Once the black holes at their centers finished sweeping up everything nearby, the quasars died. What we observe now is only their light still traveling across the Cosmos.

Many galaxies we see today may contain burned-out quasars, black holes, at their centers. Will future studies confirm astronomers' suspicions that a black hole exists in the midst of our Milky Way Galaxy?

9

Island Universes

STAND outside on a clear, cloudless night away from the glow of city lights and you will see a luminous band stretched against the blackness of the heavens. To stargazers of old, it looked as if a trail of milk had been spilled across the sky from horizon to horizon. So they called this luminous band the Milky Way.

The Milky Way, or simply, the Galaxy, is a great cluster of stars, some 200 billion of them, including the sun. The Earth is part of this Galaxy, a tiny planet orbiting the sun along with the other planets and moons of our solar system. Only a few thousand of the billions of stars that make up our Galaxy can be seen with the naked eye. Even with a powerful telescope we can see but a mere fraction of the stars in the Milky Way.

The Milky Way, like all galaxies, is moving through space. As it travels, its whirling motion flattens the Galaxy. When we look at the Milky Way we are seeing it edge-on, and it appears like a strip, or band, much as a spinning phonograph record looks if viewed side-on. If we could travel outside our Galaxy and observe it from a spaceship, however, the same edge-on view would look more like a flat disc bulging at the center where there is a great concentration of stars called the galactic nucleus. The Galaxy, in this view, looks like two orchestra cymbals stuck together face to face. The sun is located out toward the edge of the Galaxy, about three fifths of the way from the center of the disc to the rim. Although the disc is 100,000 light-years wide, it is only 2000 light-years, or one fiftieth as thick.

A spaceship traveling over the Galaxy would have a very different view of the Milky Way. Looking down from above, the Galaxy would appear to be a vast swirling spiral with several arms turning around its center. It would resemble a great cosmic pinwheel. The sun and its family of planets are located out in the far tip of one of the spiral arms. Our solar system rotates around the center of the Galaxy, making one full turn every 250 million years. Since its birth, the sun, with the Earth and its sister planets, has completed 20 such trips.

Most of the bright stars and the gas from which new stars form are found in the great trailing arms of the Galaxy. These swirling blades, however, are not uniformly dense. Gravity has gathered groups of stars together in bunches. Even so, the distances between stars are enormous. An average interstellar distance in our Galaxy is 30 trillion miles. Even Alpha Centauri, a star relatively close to our sun, is 25 trillion miles away. To get a rough idea of this distance, consider that 5000 solar systems like ours would fit between us and our closest neighboring star.

But these vast distances may be even easier to under-

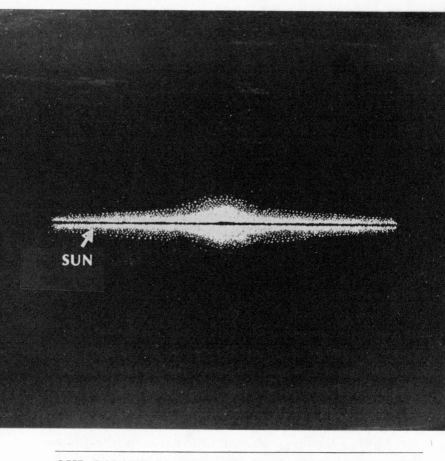

SUN

OUR GALAXY. *In this drawing, we see the shape of the Milky Way Galaxy edge-on as it would appear to an observer elsewhere in space. The Galaxy is flattened by its spinning motion; it looks like a disc with a swelling at the center. The Milky Way Galaxy is 100,000 light-years wide. Its central bulge, about 2000 light-years thick, contains the greatest concentration of stars and is therefore the brightest area. The sun (and our solar system) is located about three fifths of the way from the center to the rim of the Galaxy. (Adapted from* The Universe *by Otto Struve, published by the MIT Press, Cambridge, Massachusetts)*

stand if we reduce their scale. Think of the sun as an orange; then the Earth would be a mere grain of sand, circling it at a distance of 30 feet. Pluto, the farthest planet in our solar system, would be another grain of sand orbiting the sun 10 city blocks away. The sun's nearest neighbor, Alpha Centauri, would be another orange 1300 miles distant. On this scale the whole Milky Way becomes a bundle of 200 billion oranges that are about 2000 miles apart.

It takes extraordinary imagination to envision the magnitude of space. For the Milky Way is only one of countless "island universes," one of numberless galaxies, whirling through space. No less than 10 billion other galaxies similar in size to our own are within range of the 200-inch telescope at Mount Palomar. Enormous distances separate the galaxies. Andromeda, the nearest galaxy comparable in size and number of stars to the Milky Way, is 2 million light-years away. This is 20 times the width of the Milky Way Galaxy.

A cosmic riddle asks: Which came first, the stars or the galaxies? The current view holds that the galaxies formed before the stars within them. If the stars arose first and then gathered together later to form the galaxies, there would be no star formation going on within the galaxies today. But the opposite is true, star formation is still continuing within the galaxies now.

How then did the galaxies form? Picture a time after the Big Bang, some 15 billion years ago, when matter was evenly scattered through space. The young Cosmos consisted of one great, swirling cloud of hydrogen atoms. Now and again numbers of these particles would be accidentally swept closer together and then kept from separating once

more by gravity. These clusters of particles formed the beginning galaxies. By a similar process, star-clouds, within the forming galaxies, gathered and gave birth to the stars. When numbers of stars developed, the early galaxy formations matured into full-fledged galaxies.

Astronomers believe that for a billion years or so after the Big Bang the galaxies were evenly scattered throughout the Cosmos. Then gravity, which caused the galaxies to form, drew the nearest of them closer together into groups. Over time, these groups, each with a few galaxies, were pulled together into larger clusters containing hundreds, and even thousands, of separate galaxies. The greatest of these became superclusters. Each collection of galaxies is a hubbub of activity with individual galaxies revolving around one another like swarming bees.

Now a larger picture of the Cosmos is emerging. Great empty spaces remained when gravity drew the galaxies into groups and huge clusters. Astronomers are currently studying and mapping these voids. Future space explorers will set out into a vast cosmic sea interrupted here and there by widely separated clusters of galaxies, clusters of "island universes."

Our Milky Way belongs to a sparse cluster of galaxies called the Local Group. This cluster includes Andromeda and some 20 or so other galaxies. All are within 3 million light-years of our Galaxy. Our nearest galactic neighbors,

---

A SPIRAL GALAXY. *Viewed face-on, our Galaxy looks like a great, spinning pinwheel. It has a spiral shape like that of galaxy NGC5457 in this photograph. New stars form out in the long, trailing arms of spiral galaxies. (Palomar Observatory Photograph)*

about 200,000 light-years away, are two small, satellite galaxies held captive by the gravitational force of the billions of stars in the Milky Way. Visible in the sky of the Southern Hemisphere, they were reported by the Portuguese explorer, Ferdinand Magellan, on his famous voyage around the world (1519–1522). In his honor, they are known as the Magellanic Clouds. The Magellanic Clouds and Andromeda are the only galactic neighbors that we can see with the naked eye—they appear as faint patches of light.

But if we could soar out into the Cosmos in an intergalactic spaceship, we would see hordes of galaxies, and they would have different shapes. Most of them, including the Milky Way and Andromeda, would appear to be great spirals whirling in space. Edwin Hubble studied hundreds of galaxies and classified them by form. He divided galaxies into four main types: elliptical, spiral, barred spiral, and irregular.

Elliptical galaxies are shaped like pancakes, or flattened spheres. If you began letting the air out of an inflated

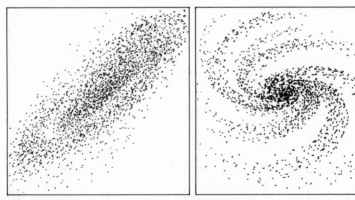

ELLIPTICAL   SPIRAL

basketball, you would produce all the many elliptical shapes that these galaxies have right down to the almost flat ones. Elliptical galaxies are the second most common type and make up about 20 percent of the galaxies. They contain stars that came into existence shortly after the Big Bang. Unlike more recently born stars, formed from the debris of supernova explosions, these old stars contain no carbon, oxygen, iron, or heavier elements. They are made up entirely of the earliest cosmic elements, hydrogen and helium.

Elliptical, spiral, and irregular galaxies (galaxies without a clearly defined shape, such as the Magellanic Clouds) are found in our Local Group—only barred spirals are absent. Barred spirals resemble the spiral type except for the presence of a prominent bar of glowing gas and dust across their center.

TYPES OF GALAXIES. *These drawings show the four main shapes of galaxies identified by Edwin Hubble.*

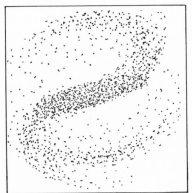 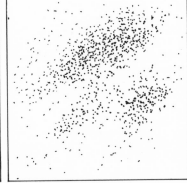

BARRED SPIRAL   IRREGULAR

What accounts for the different shapes of the galaxies? Do galaxies always keep the same form or do they change shape over time? Some astronomers suspect that, within a given cluster, galaxies may collide, gobble one another up, or even pass right through one another. Collisions between galaxies might result in new formations or lead to combined galaxies of different shapes. But such events could occur without individual stars ever bumping. Collisions between stars are rare because the average distance between stars is so great in comparison to the size of the average star. For example, in the Milky Way Galaxy, the average distance between stars, about 30 trillion miles, is millions of times greater than the size of a star. Astronomers know far more about the life cycle of stars than about the evolution of galaxies. Many questions about the galaxies remain unanswered.

10

FROM DARK ORIGINS:
EARTH

IMAGINE our Galaxy as it was almost 5 billion years ago. The birth and death of stars had been going on for billions of years by now, and many stars lit up the heavens. The sun, the Earth, and its sister planets, our solar system, however, did not yet exist. The region where they would appear was still a cold, dark corner of the Milky Way.

In this void, out near the end of one of the Galaxy's great, spiral arms, a mist of gases and cosmic dust was stirring. This mist consisted largely of the light gases, hydrogen and helium, the most common elements in the Cosmos. Mixed among the hydrogen and helium were atoms of all the common elements and some of the rarer ones, too. These substances had been manufactured in the nuclear ovens of great stars and blasted out into interstellar space by supernova explosions.

Over time, gravity gathered this thin mixture of stardust and gases closer, and a dark, swirling cloud formed from the mist. As the density of this cloud increased, its gravity became stronger. Atoms raced inward toward the center of the cloud; the process of star formation began. For millions of years, this process continued: Matter accumulated in the contracting center of the cloud, the core grew hotter, nuclear reactions took place, and finally the hydrogen ignited and burst into flame. The sun, the star at the center of our solar system, was born. It brought light and heat to this once dark corner of the Galaxy.

Astronomers are less certain how the planets of our solar system evolved. They believe that the process began slowly, however, and then quickened in the later stages. While the center of the parent cloud was contracting to form the young sun, other changes were taking place out at the edges of the cloud. Here, where the mist was cooler and less dense, the mixture of gases and stardust formed a halo around the shrinking core. The particles of matter in this halo circled the contracting center of the cloud,

---

THE FORMATION OF THE SUN AND PLANETS. *Figure 1: Almost 5 billion years ago, out in one arm of the Milky Way Galaxy, a great cloud of gas formed with a radius of about 10 trillion miles. It was held together by the force of its own gravity. Figure 2: Gradually, as gravity drew the atoms of gas together, this diffuse cloud contracted. Figure 3: With the passage of still more time, perhaps 10–20 million years, the dense center of the cloud shrank to the size of the present sun. At this point the temperature at the core of the contracted cloud had risen high enough to ignite nuclear reactions; hydrogen atoms fused to form helium, releasing energy and marking the birth of the sun. Figure 4: In the cooler, less dense regions surrounding this new star, the sun, smaller condensations developed, giving rise to the planets.* (Red Giants and White Dwarfs *by Robert Jastrow, published by Warner Books*)

1

2

3

4

MARS

VENUS

MERCURY

SUN

EARTH

colliding with one another as they spun. These accidental collisions were less violent than those at the center, in the forming star. Nuclear reactions did not occur out here. Instead, the colliding atoms in the halo often stuck together, creating larger particles.

We know all the natural mineral substances and many common chemical compounds were formed in this halo around the embryonic sun. There were grains of pure iron and bits of rock made from iron, silicon, oxygen, aluminum, and magnesium. There was water vapor from hydrogen and oxygen, ammonia gas from hydrogen and nitrogen, methane (or marsh gas) from hydrogen and carbon, and carbon dioxide from carbon and oxygen—and countless other combinations of elements as well.

The particles in the surrounding halo continued to collide, stick together, and grow larger as they circled the ever-shrinking center of the cloud, with its ever-increasing temperature. The temperature in the halo, however, remained extremely cool, perhaps 100 degrees Fahrenheit below zero. Water vapor froze into tiny ice crystals that mixed with ammonia and methane to form a slushy mixture. Carbon dioxide gas solidified into dry ice. Gradually, the halo changed from an envelope of fine, microscopic particles into a zone of coarse, growing fragments.

This phenomenon, the joining together of tiny particles to form fewer, larger bodies, is called condensation. Early on, it occurred more or less accidentally. But as larger masses of ice crystals, grains of iron, and bits of rock formed, they began to exert a gravitational attraction on neighboring matter. The condensation process speeded up, and larger fragments swept up everything nearby.

Eventually, these larger bodies changed position. Like raindrops condensing from a mist, they began to "fall." They fell from the halo down to the level where the sun's gravity was strongest and the orbiting gases were most dense. In this plane, they continued to attract additional matter. These solid, growing masses were the seeds of the planets.

It took millions of years for the planets to reach their present size. By then, condensation of the cooler, outer reaches of the parent cloud had produced nine planets, several moons, and numbers of smaller bodies called asteroids and comets. Asteroids are solid, rocky masses that circle the sun in a belt between the orbits of the planets Mars and Jupiter. These objects range in size from small stones to bodies rarely greater than a few miles in diameter. They are sometimes called planetesimals, or small planets. No one knows if these masses are fragments of a planet that broke apart, or matter that, for some reason, never came together to form a planet.

Comets are masses of frozen gas and dust that circle the sun in very wide orbits. We see them from Earth only rarely. At such times, when their travels bring them closer to the sun, its heat vaporizes their gases and causes them to glow. A comet looks like a luminous ball with a long streaming tail.

The solar system was completed about 4.6 billion years ago. Rocks brought back to Earth from the moon by the *Apollo* astronauts who landed there in 1969 helped to confirm the age of the solar system. The moon has no atmosphere, no winds or oceans to erode its surface, so objects there have lain relatively undisturbed since the

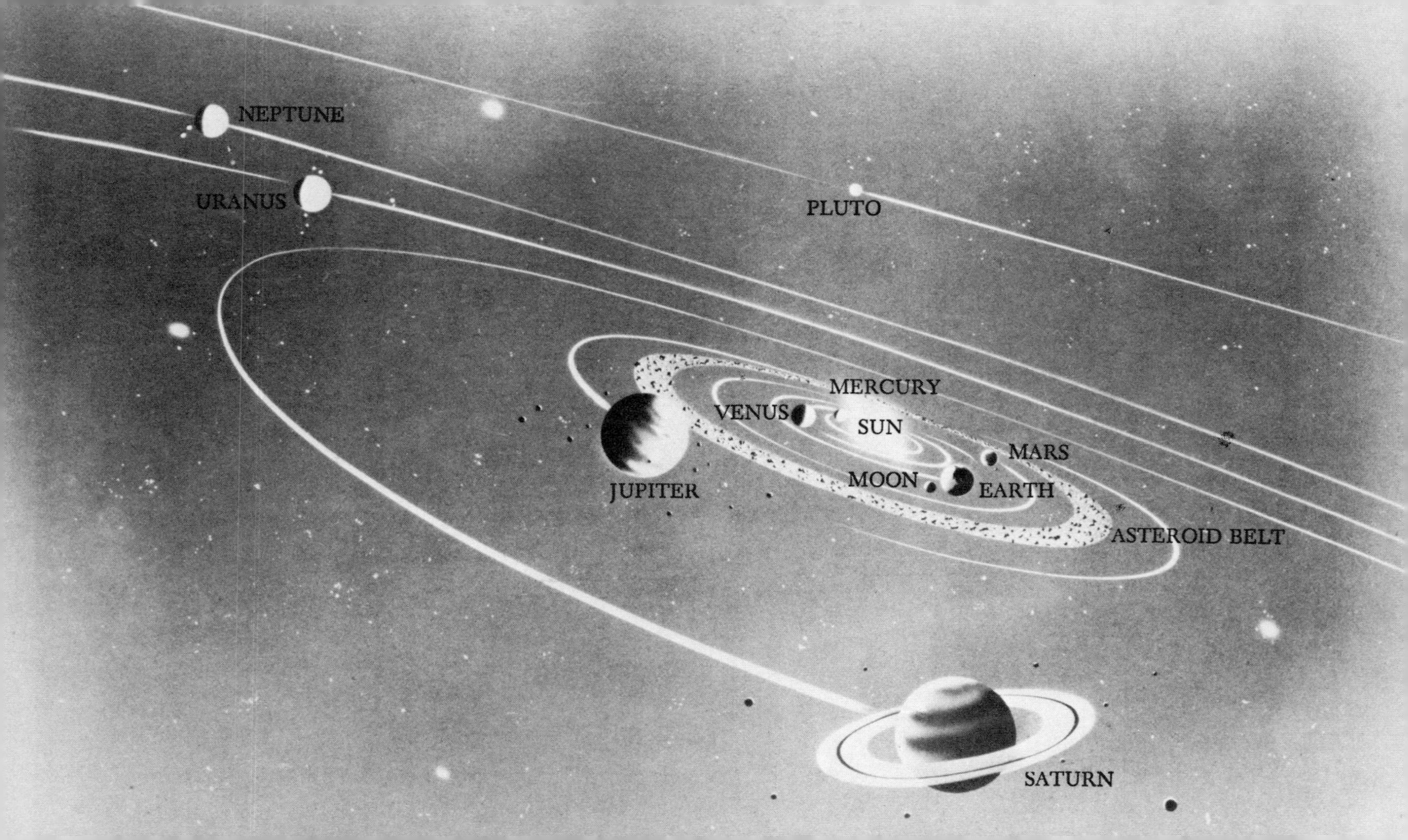
NEPTUNE
URANUS
PLUTO
MERCURY
VENUS
SUN
MARS
MOON
EARTH
JUPITER
ASTEROID BELT
SATURN

moon's formation. None of the moon rocks, which have been studied and analyzed intensively, were older than 4.6 billion years.

Our understanding of the solar system's evolution is far from complete, however. Astronomers are not certain, for example, which was finished first, the sun or the planets. It may seem strange, but it has been easier to learn the secrets of stars trillions of miles away than to piece together the history of our own Earth and its sister planets. The light of the stars tells their temperature, composition, and much information about their nature. The planets we observe, on the other hand, shine only by reflected sunlight from which we learn relatively little. Astronomers can also study stars in all stages of development, but planets of different ages have not been found. So great mysteries remain about their formation.

For example, what became of the hydrogen and helium that were so abundant in the parent cloud of the solar system? You might expect that these gases would make up much of the mass of all the planets, but this is not so. Only the outer planets Jupiter, Saturn, Uranus, and Neptune,

---

THE SOLAR SYSTEM. *Nine planets revolve in orbits around the sun bound to it by gravity; the weight of the sun is 700 times greater than the combined weight of the nine planets. The sun, the planets, their satellite moons, and a number of other smaller bodies including asteroids and comets form the solar system. The width of the solar system is 8 billion miles across. An exceedingly sparse cloud of hydrogen gas fills the nearly empty space between the planets and the space beyond the solar system. (Adapted from the* Reader's Digest Great World Atlas, *by permission of the* Reader's Digest Association)

THE LUNAR HIGHLANDS. *The lunar highlands are considered to be the oldest area on the face of the moon. In this photograph, geologist Jack Schmitt of the Apollo 17 mission (December, 1972) examines a huge boulder that was probably dislodged from the lunar surface by the impact of a large meteorite. Smaller moon rocks brought back to Earth from this area have been determined to be 4.6 billion years old, as old as the solar system itself. (National Aeronautics and Space Administration)*

the largest planets in the solar system, are composed mainly of hydrogen and helium. The Earth and the other planets closest to the sun have little of these light gases. Why? Were these gases driven off by the heat of the sun—hotter back then than it is today? Or did the opposite occur? Were they drawn into the young, forming sun as it contracted? Whatever the case, the fact remains that these light gases largely disappeared from the Earth and the other innermost planets by the time they were completed.

The Earth, when it first appeared, was a solid mass without a gaseous atmosphere. It was a dry, barren chunk of rock created from bits of material that had come together by accident and by gravity over millions of years. These fragments were coarse and irregular. When they joined, gases and other substances—including rare and heavy elements—were trapped inside the forming planet. Certain of these elements were to play a crucial role in the evolution of the Earth.

For some of the rare elements buried deep within the Earth have unique qualities. They are radioactive. They break down by themselves, emitting particles of their atoms as they decay. These elements, such as uranium, thorium, and others, have been disintegrating in the interior of the planet ever since it first formed. The particles they emit smash into atoms of the surrounding rocks and heat them. Gradually, over hundreds of millions of years, this radioactivity in the interior of the Earth caused the solid center of the planet to melt. Hot, liquid rock filled the core of the Earth.

The melted material expanded, but it was trapped inside the planet. It rose to fill cracks and spaces and to push against weak areas in the Earth's surface. In a number of places, it burst through. Volcanoes erupted, spewing forth torrents of molten rock. Along with this hot lava, trapped gases were released, and water vapor, in the form of steam, bubbled out.

Volcanic activity increased, and, with each new, violent eruption, additional water vapor and trapped gases escaped with the lava. Thick clouds formed above the erupting volcanoes. As they cooled and condensed, warm rains fell. This moisture streamed over the naked surface of the Earth, filling depressions and spreading until a shallow ocean covered the entire planet. Still more volcanoes erupted, heaping up mounds of congealed lava until islands formed and rose from the bottom of the sea. Meanwhile, above the changing surface of the planet, certain gases did not condense. These gases formed the Earth's first atmosphere; this air was composed largely of nitrogen.

So the molten interior of the planet gave rise to the oceans, land, and atmosphere. The Earth, at first a dry, barren, inhospitable lump of rock, gradually changed into a planet ready for life.

When plant cells eventually evolved, they produced oxygen as a waste product and discharged it into the atmosphere. The greening of the planet added substantial amounts of oxygen to the air for the first time. The presence of oxygen created conditions necessary for the development of still higher forms of life.

11

The Innermost Planets: Venus and Mercury

T HE Age of Exploration here on Earth began in the year 1492, when three little ships sailed from Spain on a voyage into the unknown. Not even their daring commander, Christopher Columbus, could have dreamed that the *Niña, Pinta,* and *Santa María* would affect the lives of all peoples thereafter.

Today, a new Age of Exploration has begun: the Exploration of Space. Names like *Mariner, Viking, Pioneer,* and *Voyager* make history now. For these small, fragile spacecraft also journey into the unknown, seeking answers to the many mysteries of our solar system. And the men and women who direct these missions are also finding wondrous New Worlds out in space. Each unmanned space probe changes our ideas about the other planets. In fact, the more we learn, the more we realize how little we know about our neighbors in the solar system.

Exploration of the other planets began in 1962. Five years earlier, *Sputnik 1,* the first artificial satellite, had been launched by the Soviet Union; it opened the Space Age. But the first visit to another planet was an American mission. It was planned and directed by scientists at the National Aeronautics and Space Administration (NASA). The craft was named *Mariner 2*, but it hardly resembled any ship ever launched upon our seas. *Mariner 2* looked more like a bug. Its broad, flat solar panels resembled stubby wings, and its long radio antennas looked very much like feelers.

Mercury is the first planet from the sun, Venus is the second, and Earth is the third. *Mariner 2* was to visit Venus. Venus is similar in size and mass to Earth, so it was always considered to be Earth's twin. It is the nearest planet to Earth. But thick clouds hide Venus' surface from view, so we have never actually seen it. This dense cloud cover reflects the sun's rays and helps to make Venus one of the brightest objects in the sky. But what secrets do the clouds of Venus conceal?

Could Venus harbor life? Venus is 67 million miles from the sun—the Earth is only 26 million miles farther. This difference is relatively small considering the vast distances in space. Venus receives twice as much sunlight as the Earth, but astronomers still thought it might enjoy a comfortable climate. Because of the planet's protective cloud cover, some believed Venus would have a warm, pleasant climate, like that of a tropical island here on Earth. Venus, named for the Roman goddess of beauty, was pictured as a lush paradise covered with plants and teeming with life. The French scholar, Bernard de Fontenelle, wrote in 1686:

> I can tell from here . . . what the inhabitants of Venus are like; they resemble the Moors of Granada;

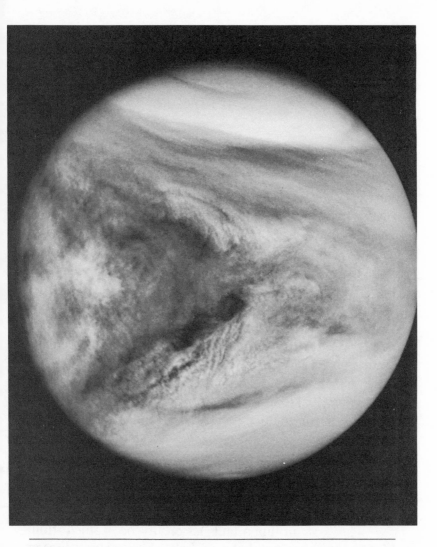

VENUS. *Because it is similar in size, weight, and distance from the sun (Venus is only 26 million miles closer), Venus is considered Earth's sister planet. The face of Venus has always been hidden from view beneath a dense blanket of clouds, leading to fanciful notions that Venus might enjoy a balmy, tropical climate. Radio astronomers learned otherwise. (National Aeronautics and Space Administration)*

a small black people burned by the sun, full of wit
and fire, always in love, arranging festivals, dances
and tournaments every day.

What a disappointment it was, therefore, in 1955, 7
years before the *Mariner 2* launching, to learn that the surface of Venus was like an oven. Radio telescopes trained
on Venus penetrated its clouds and revealed the planet's
temperature to be hot enough to melt lead. In December,
1962, when *Mariner 2* visited Venus, it confirmed that the
planet's temperature was sizzling—800 degrees Fahrenheit.
No known form of life could survive such heat.

Since the first *Mariner* mission, there have been more
than a dozen visits to Venus by American and Russian
spacecraft. And unmanned Russian probes have actually
landed on its surface. Spacecraft, armed with cameras and
other scientific equipment, show the planet to lack many
factors essential to life. Venus turns out to be almost waterless except for some superheated steam. Its fluffy white
clouds consist of droplets of deadly sulfuric acid. The
surface of Venus is bathed in poisonous carbon dioxide—
the gas is so dense that it exerts a crushing pressure one
hundred times greater than the atmosphere on Earth. Venus is a hot, barren, hellish place.

There is an important lesson to be learned from
Venus, however, one that may help us to preserve life on
our own planet. For the carbon dioxide that covers Venus
lets sunlight through but keeps warmth from escaping. In
other words, it acts like the glass of a greenhouse. The result of this "greenhouse effect," as it is called, is searing
heat and desolation on Venus. It is a warning to us here

on Earth. If we overuse fuels that produce carbon dioxide, such as coal, oil, and natural gas, we could cause a dangerous change in our atmosphere. Careless production of too much carbon dioxide might create a "greenhouse effect" on Earth.

But what is hidden beneath the acid clouds covering Venus? Radar, radio waves beamed at Venus from Earth and recorded as they bounce back to us, show the face of Venus has craters, canyons, plateaus (high flat areas), and a mountain range higher than any on Earth. But overall, the surface of Venus seems unremarkable and fairly smooth.

The strange thing about Venus is the way it rotates. Venus spins backwards, turning clockwise on its axis. Venus rotates from east to west as it revolves around the sun. All the other planets in our solar system rotate from west to east, or counterclockwise. Only on Venus does the sun rise in the west and set in the east. Venus also rotates very slowly. A day on Earth, 24 hours, is the time it takes our planet to make one full turn on its axis. A similar rotation of Venus takes 243 Earth days. Venus has another distinction. It is the only planet, except for Mercury, to have no natural satellites.

Mercury, less than half the size of the Earth, is the innermost planet in the solar system. It is barely larger than our moon. Mercury is only 36 million miles from the sun. Since it is the closest planet to the sun, Mercury has the smallest orbit. It completes one full trip around the sun in only 88 days, making Mercury the fastest planet in the solar system. This rapid movement accounts for its name. Mercury was the speedy messenger god of ancient Rome.

For many years, astronomers believed that Mercury did not rotate at all as it traveled around the sun. They thought that the planet always kept the same side to the sun. It was believed that side had scorching, endless day while the other side, facing away from the sun, had frigid, everlasting night. But, in 1965, astronomers, using radio telescopes and radar, discovered that Mercury does rotate. It spins very slowly, taking 59 Earth days to complete each turn.

Because Mercury is so near the sun, it is difficult to see. It gets lost in the sun's brightness. This glare also ruins photographs of the planet, causing Mercury to look like a fuzzy ball of cotton. We can only see Mercury for a short time. It appears near the horizon just before sunrise and just after sunset. Unfortunately, that is where the haze in the Earth's atmosphere is worst, so we only get a brief, blurry view of Mercury even though it is relatively close to Earth.

The first clear pictures of Mercury were sent back to Earth from the spacecraft *Mariner 10* in 1974. These photographs showed the planet to have craters, ridges, mountains, valleys, and great smooth plains of lava, formed by volcanoes now long extinct. A distinctive feature of Mercury, however, are huge cliffs that stretch for hundreds of miles across its face. These cliffs, or rupes, as they are called, are 2 miles high in places. Astronomers are honor-

CRATERS ON THE MOON. *Craters in the surface of the moon and the inner planets provide important clues to the history of these bodies. This photograph of craters on the moon was taken through the 100-inch telescope at Mount Wilson. (Mount Wilson Observatory Photograph)*

ing the brave, little explorer ships of yesteryear by naming the rupes after them. One line of rupes has been named the *Santa María* Rupes, after Christopher Columbus's flagship.

Except for the rupes, the cratered, dusty surface of Mercury generally resembles that of the moon. It is waterless and all but without an atmosphere. Mercury is a sunscalded hunk of barren rock. It could not, by any stretch of the imagination, ever support life.

12

The Red Planet: Mars

F life exists on any other planet in the solar system
besides Earth, then Mars might well be the place to
find it. Indeed, for years, science-fiction writers have told
exciting stories of superintelligent, odd-looking creatures
who live in advanced civilizations on the red planet, Mars.
On one occasion, a dramatic radio account of a Martian
invasion of Earth was so convincing that listeners, believ-
ing it to be true, were frightened and panicked. But the
possibility that life might actually exist on Mars was not
limited to science fiction. Respected astronomers also believed that there was a real chance of finding advanced
forms of life there.

In 1887, the Italian observer, Giovanni Schiaparelli,
claimed to see a vast network of "canali" reaching far and
wide across the face of Mars. Then others, including the
distinguished American astronomer, Percival Lowell, reported seeing the canals too. Surely, they claimed, the
channels could only be the handiwork of intelligent beings

with extraordinary engineering skills. Sketches of the Martian canals aroused much public interest. But not all astronomers agreed that the canals existed, and interest in them faded when photographs failed to confirm their presence.

Still other factors, however, seemed to hint at life on Mars. Four times each Martian year (687 Earth days), for example, the surface of the planet changes in a way that suggests that Mars has seasons like those on Earth. In each hemisphere, dark regions appear every spring as though areas of the planet are acquiring fresh growth. The polar ice caps in each hemisphere also vary with the seasons, growing larger in fall and winter and shrinking in spring and summer.

These factors suggest that water is present on Mars. Water, scientists agree, is the one substance that must be present and plentiful for almost any kind of life to begin. It is the essential medium through which minute bits of matter can drift, collide, and join together to evolve into living organisms. Astronomers hoped to find water on Mars.

Mars is 142 million miles from the sun, half again as far from the sun as Earth. It is the fourth planet from the center of the solar system. Mars is about midway in size between the moon and the Earth. It has two tiny moons and is surrounded by a thin atmosphere with only a trace of clouds. Without dense clouds to conceal its surface, one might expect that any mysteries about Mars would have been solved a long time ago. But this is not the case.

Currents in the atmosphere of our own planet prevent a clear view of the Martian landscape. In fact, it is impossible to get good photographs of the surface features of Mars from Earth even through the largest telescopes. So observers could not tell from here whether the face of Mars is smooth or cratered, or, most important of all, whether volcanoes exist there. For if volcanoes formed during that planet's development, as they did on Earth, then the chances of finding water and an atmosphere favorable to life there would be greatly improved.

In 1965, NASA launched *Mariner 4* to take photographs of Mars. But the pictures the spacecraft sent back proved to be disappointing. They were grainy and generally of poor quality. *Mariner 4* was only able to photograph a section of Mars, and the pictures it took failed to show volcanoes or the presence of free bodies of water—lakes or oceans. There was nothing to encourage hopes of finding life.

The pictures did reveal, however, that the surface of Mars is cratered. The craters are evidence that, like the moon, the Earth, and the other inner planets of the solar system, Mars, too, has been bombarded with meteorites. These fragments of rock and iron, which range in size from tiny grains to great masses weighing a million or more tons, are believed to come from the asteroid belt between Mars and Jupiter.

The craters of Mars provided clues to the planet's physical history. Scientists believe that the rain of meteorites that pitted the moon and the inner planets took place largely during the first 500 million years after the solar system formed. Therefore, the number and condition of the craters on these bodies show differences in their physical development since then. Our moon, for example, has many well-preserved craters, evidence that there has been little action to erode the surface there. The Earth, on the other hand, has few craters. Wind, running water, volcanic activity, and shifts in the Earth's crust have erased most of them. Photographs of the craters on Mars showed them to be older and better preserved than those on Earth, but not as intact as those on the moon. Some kind of surface activity in the past has taken place on Mars.

So, despite disappointment that the first pictures from *Mariner 4* revealed Mars to be a barren, lifeless place, the evidence of erosion kept up some slim hope of finding life there. NASA planned another mission to the planet. In May, 1971, it launched *Mariner 9*, a spacecraft equipped with improved cameras and instruments. But, unfortunately, when *Mariner 9* approached Mars and went into orbit around the planet, the surface of Mars was hidden. The planet was experiencing a violent dust storm. Local dust storms occur frequently on Mars, but this one was especially turbulent and covered the surface of the entire planet. The storm lasted for 2 months. So, after all their plans and preparations, scientists at the Jet Propulsion Laboratory in Pasadena, California, had to wait patiently for the storm to end and the dust to settle.

Not until early 1972 did *Mariner 9* begin to send back clear photographs of Mars. Now the patience of those responsible for this mission was rewarded. The spacecraft orbited and photographed the whole planet. Its pictures revealed wonders never suspected from the earlier Mariner photographs. Mars has spectacular features. One of these is a great rift that splits the surface of the planet at its equator. This extraordinary rift drops to a depth of 20,000 feet and is 150 miles wide in places. Most remarkable of

A MARTIAN LANDSCAPE. *This photograph, taken by a Viking lander in 1976, shows an area on the Plain of Chryse in the northern hemisphere of the red planet, Mars. The desolate scene resembles desert regions of the southwestern United States. The windblown soil is rust colored, evidence of its rich iron content. Rocks strewn about the surface are bits of cooled molten lava that came from active volcanoes. The large rock at left is about 3 feet high and 25 feet from the camera. (National Aeronautics and Space Administration)*

127

all, the rift extends for some 3000 miles, about the width of the United States from the East Coast to the West.

Even more noticeable, however, were a number of huge, extinct volcanoes. The greatest of these rises 15 miles high from a base that is 300 miles across. This mammoth volcano, which has been named Mount Olympus, dwarfs the largest volcanoes on Earth.

The discovery of volcanoes on Mars was the most exciting find of all to those who still hoped to find life on the planet. For although Mars appears relatively dry today, the volcanoes indicate that, some time ago, it may have had plenty of water, which is essential for the existence of life as we know it. There were also other clues to suggest that water once existed on Mars. *Mariner* 9 photographs showed curving channels, some, hundreds of miles long, that unmistakably resemble dry riverbeds with tributaries. These channels are probably millions, or billions, of years old. They appear to be the type that flash floods, carrying hundreds of tons of rapidly rushing water, would carve into the face of a planet.

But where is the water that created these ancient riverbeds? *Mariner* 9 found no open bodies of water on Mars. Did life, if it ever existed on Mars, disappear with the water, or could living organisms have adapted to the drier conditions and somehow managed to survive?

With this question in mind, NASA sent two additional spacecraft, *Vikings* 1 and 2, to Mars. In the summer of 1976, the *Viking* spacecraft orbited Mars and took photographs and measurements at its northern pole. Scientists had come to believe that the polar ice caps of Mars were mainly made of dry ice, frozen carbon dioxide. That summer, to their surprise, Viking showed the north polar cap to be composed of melting frozen water. The photographs also revealed deep craters filled with ice beneath the polar cap.

So at last, there was an answer to the puzzling whereabouts of the moisture freed from the interior of Mars in earlier volcanic eruptions: It is still there, trapped in the polar ice caps and in great ice-filled craters in the ground beneath them. If all the ice that is estimated to be in the poles and craters of Mars were melted, water could cover the entire planet with a shallow sea.

The spacecraft's photographs also revealed telltale layering of the polar ice. This shows that the thick polar caps had melted during warmer periods in the past and refrozen. So Mars was warmer and wetter in the past but is in the midst of an ice age now. The current ice age has locked in the moisture, turning the surface of the planet into a desert. Yet water could not exist in liquid form on Mars today. The atmosphere of Mars is too cold and too thin now. The air there is only one hundredth as dense as that of the Earth, so the sun's rays would evaporate liquid water quickly.

But if water had existed in abundance during warmer periods, hundreds of millions of years ago, simple forms of life might have evolved and taken hold. Then, over time, as the planet cooled and became more desertlike, organisms, better adapted to these harsh conditions, might have survived and reproduced. Such life would be simple—perhaps primitive organisms such as bacteria. Certainly, Martians would not be complex, thinking creatures. Scien-

THE VIKING LANDER. A *marvel of modern space technology, the Viking lander is shown here in a rehearsal exercise at the Jet Propulsion Laboratory in Pasadena, California. In a remarkable feat, the Viking lander was set down on the planet Mars, within a few miles of its target, after journeying 200 million miles.*

The primary mission of the lander was to test for the presence of life on the red planet. In performing its function, the lander was guided by an onboard computer able to receive commands from its base at the Jet Propulsion Laboratory. Its equipment included a retractable arm (foreground) for collecting samples of soil, and biological and chemical processors which received these samples and analyzed them for signs of life. The lander also carried a seismometer to register earthquake activity, a small weather station, cameras, antennas—and the hopes and prayers of hundreds of curious scientists. (National Aeronautics and Space Administration)

tists estimate that conditions favorable to the development of advanced forms of life did not last nearly long enough for that to take place.

The second phase of the *Vikings'* mission involved setting down two probing devices on the red plains of the planet. Each spacecraft carried a marvelous, completely automated lander, containing cameras, a miniature laboratory, and an ingenious robot arm. These probes were ejected and gently landed in different regions. On command, the landers' robot arms scooped up the dusty, red soil of the planet and fed the samples into the laboratory chambers for analysis. The Martian soil is red, like the earth of the southwestern United States, because it is rich in iron. The material was tested for biological activity.

The tests were designed to reveal the presence of life, no matter how simple its form. The first probe reported test results that could just as well have come from chemical reactions as from biological activity. The second probe, 4600 miles away, however, reported findings somewhat more likely to have a biological explanation.

The question of life on Mars may not be settled for many years to come. Nevertheless, the *Viking* experiments, even with their contradictory results, hint that life of some kind may exist on Mars.

13

Giant Planets

BEYOND Mars lie the five remaining planets of our solar system: the giant planets, Jupiter, Saturn, Uranus, and Neptune, and the small planet, Pluto. Only recently, by means of unmanned spacecraft, have we been able to learn many of the outer planets' secrets. But there is still much mystery about them.

Pluto, the ninth and outermost planet, was only discovered in 1930. We still know little about it because it is so distant and has yet to be explored by spacecraft. Pluto is the smallest planet in the solar system; it is only slightly larger than our moon. Pluto appears to be a frozen world quite unlike its huge, gaseous neighbors. Indeed, small Pluto seem misplaced at the fringe of the giant, outermost planets. Its presence there with its single moon, Charon, is a riddle.

The giant, outermost planets are about 10 times larger in size than the Earth and have much more mass. They contain about 100 times more matter. This matter, however, is less tightly packed together—most of it is in the form of gas. These planets are composed mainly of the light gases, hydrogen and helium. In fact, Saturn and Jupiter resemble the sun in makeup more than the Earth. But Jupiter, the largest planet in the solar system, has only one thousandth the mass of the sun. It lacks sufficient mass to generate the nuclear reactions needed for it to become a star. Nevertheless, Jupiter radiates more heat into space than it receives from the sun. This excess heat may be energy left over from the original formation of the planet 4.6 billion years ago.

Jupiter is the fifth planet from the sun. It is named for the powerful ruler of the gods of ancient Rome. This is fitting because Jupiter is larger than all the other planets of the solar system combined. It is a gigantic, rapidly spinning ball of gas and liquid about 318 times more massive than the Earth. It takes Jupiter 12 Earth years to make one full trip around the sun because its orbit is much larger than that of the Earth. But it rotates so quickly that a full day and night on Jupiter takes less than 10 Earth hours.

The massiveness of Jupiter gives it far greater gravity than our planet. In theory, an astronaut approaching the surface of Jupiter to land would find his/her weight increased to about 600 pounds, and the person would be stuck to the spot where he/she set down. In reality, however, Jupiter has no solid surface on which an astronaut, or a spaceship, could land.

Buried deep within its interior, Jupiter may have a hard core of rock and iron. But a layer of liquid hydrogen —hydrogen gas compressed and heated until it becomes a liquid metal—surrounds this core, extending outward for some 40,000 miles. Near the cooler surface of the planet, the liquid hydrogen merges into a dense, gaseous atmosphere. Small amounts of other gases, ammonia and methane, are also present in this atmosphere. Water is there, too, as steam in the lower atmosphere where the temperature is high, and as droplets of liquid water at cooler, higher altitudes.

These droplets of water in the upper atmosphere of Jupiter form thick clouds over the planet. The water clouds and water ice, in turn, are covered by clouds composed of frozen crystals of ammonia. What we have, in effect, is an atmosphere made up of decks of clouds like the layers in a cake. In fact, the outermost layer of ammonia crystals is rather like the icing, it gives Jupiter a colorful face. The ammonia clouds appear as wide and narrow bands of yellow, red, tan, gray, and white that cover the planet.

Striking photographs of the multicolored bands of Jupiter have been taken by spacecraft cameras. Spacecraft flew by Jupiter in 1973 and 1974. *Pioneer 10* passed Jupiter on a course that took it out beyond the solar system, and *Pioneer 11* went on to Saturn. Then, in 1979, two other spacecraft, *Voyagers 1* and 2, visited Jupiter to make more detailed studies of the planet. Photographs from the *Voyager* missions revealed vast swirling storms thousands of miles wide.

The *Voyager* spacecraft also photographed Jupiter's most intriguing feature, its famous Great Red Spot. Located on its Southern Hemisphere, this huge, oval, orange-red area is 15,000 miles long and almost 8,000 miles wide, roughly the size of the Pacific Ocean. The Great Red Spot of Jupiter has been observed by astronomers for at least 300 years and has been the subject of discussion and wonder.

Now remarkable photographs by *Voyager* cameras revealed the area of the Great Red Spot to be a mammoth storm. It is the site of a hurricane of incredible size and duration—something beyond the scale of anything known on Earth. We don't know what force generates and maintains this extraordinary phenomenon—a hurricane that has persisted for at least 300 years in the same area.

As the *Voyager* spacecraft swept by Jupiter, their cameras made other discoveries. One major surprise is that Jupiter is surrounded by a ring. Scientists had wondered for a long time why, of all the planets, only Saturn seemed to have rings. Then, in 1977, observers on Earth found rings around Uranus. Now *Voyager* photographs revealed that Jupiter, too, is ringed. Jupiter's ring is too thin and faint to be seen through telescopes on Earth. It is made up of tiny, orbiting fragments of rock that circle the planet in a band 160,000 miles in diameter.

Beyond Jupiter's ring lie the planet's 14 moons. The largest of these satellites, Io, Europa, Ganymede, and Callisto, are named for Jupiter's four lovers in Roman mythology. Their rapid movements around Jupiter suggested to Galileo, in 1610, that Jupiter and its moons were themselves a miniature solar system. This was the strongest evidence at that time in support of the Copernican view that the sun, not the Earth, was the center of the Cosmos. (Of course we now know that the Cosmos has no center.) Jupiter's orbiting moons proved that all motion in the world did not revolve around the Earth. So the moons of Jupiter played a historic role, and they were also di-

rectly responsible for Galileo's troubles with the Roman Catholic Church. For religious authorities felt that Galileo's contribution to the Copernican theory lessened the importance of the Earth, and they held that to be an insult to God.

The *Voyager* spacecraft photographed the four largest moons of Jupiter in detail. Their pictures showed Callisto and Ganymede, moons similar in size to the planet Mercury, to be essentially large drops of liquid water with frozen crusts of ice and rock. Ganymede is the largest known moon in the solar system. Io and Europa, both about the size of our moon, are made up mostly of solid, rocky substances.

Io, the innermost of the four largest moons of Jupiter, provided the second major surprise of the mission. It was discovered to have active volcanoes like those here on Earth. Io is geologically alive. In fact, *Voyager* cameras captured one of its volcanoes erupting. This surprising discovery of active volcanoes on Io means that, besides Earth, Io is the only known body in the solar system to be volcanically alive. Our moon is geologically dead, and so is the planet Mars—its last volcano died out a billion

---

JUPITER. The largest of the giant planets, Jupiter, is a rapidly spinning ball of liquid and gas. On the planet's southern hemisphere is its Great Red Spot, seen here as a dark oval at the top center of the photograph. This striking feature, about the size of the Pacific Ocean, was observed by Galileo through his telescope. It is a gigantic storm that has persisted for at least 300 years. Immediately beneath the Great Red Spot is a white oval, another stormy region. (National Aeronutics and Space Administration)

years ago. Does this suggest, then, that Io has water and warmth and could support life? No. Io has no atmosphere and lacks the warm shallow seas required for the processes that lead to life.

Jupiter itself may shed some light on the puzzling question of how life begins. Jupiter's stormy, dense atmosphere with abundant amounts of ammonia, methane, hydrogen, and water contains the essential elements required for life. When these materials are mixed together in the laboratory and activated by electric charges, they combine to form substances called amino acids and nucleotides. These large molecules are the basic building blocks of life.

Scientists believe that when the Earth was a young planet, its atmosphere also contained large amounts of ammonia, methane, hydrogen, and water. They think that natural charges of electricity, bolts of lightning, energized these gases to produce amino acids and nucleotides. These molecules collected in the shallow ocean that covered the young planet at that time. The sea became a rich nutrient soup.

Over millions of years, accidental collisions between these molecules in the sea caused still larger and more varied nucleotides to form. Finally, one long molecule composed of nucleotides evolved that had the unique ability to attract similar parts and assemble copies of itself. It was also able to divide, like the two halves of a zipper. This particular molecule called deoxyribonucleic acid, or DNA for short, had acquired the marvelous ability to reproduce itself. It had crossed the line that separates the living from the nonliving. Life had begun.

These primitive bits of life came to predominate in

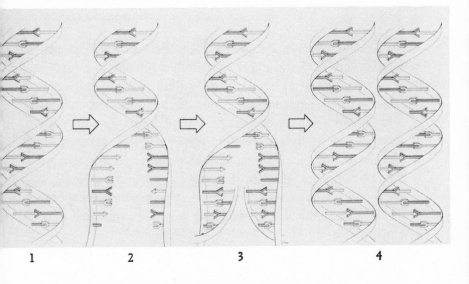

THE MECHANISM OF REPRODUCTION. *Figure 1: DNA, the master molecule of reproduction, resembles a twisted chain. The links that form this chain are made up of four types of nucleotides. These are represented here by the four symbols* ⇒ ⇒ ⇒ ⇒. *Figure 2: When a cell is about to divide, the chain untwists, each link breaks at the midpoint, and new nucleotides, as shown in Figure 3, are collected from the surrounding molecules in the cell. Figure 4: The result is two DNA molecules, replicas of the original DNA. (Astronomy: Fundamentals and Frontiers by Robert Jastrow and Malcolm Thompson, published by John Wiley and Sons)*

---

the shallow sea that covered the Earth. As they bumped, collided, and reproduced, accidents occurred that resulted in countless new formations. Each one was a natural experiment: Some worked, others did not. After a great length of time, several billion years, the successful experiments developed and evolved into the great variety of plants and animals that now exist on Earth.

So far, attempts to produce life in the laboratory have failed. It would require virtually a miracle to create, in a short time, what took nature millions of years. Jupiter, however, is a cosmic laboratory where these experiments have been going on for over 4 billion years already. *Voyager* photographs show that stormy Jupiter, rich in the elements required for life, probably has lightning as well. Amino acids and nucleotides, the basic building blocks of life, may well exist in abundance in Jupiter's atmosphere. Have these already joined and evolved into living organisms?

The chances of finding living organisms on Jupiter are slim. Jupiter has only water clouds to encourage the necessary collisions between the bits of matter that lead to life. The opportunities for such lucky encounters between particles floating in Jupiter's clouds are less favorable than would be the case if Jupiter had oceans. In all likelihood, it would take infinitely longer for life to evolve on Jupiter than it did on Earth. Still, Jupiter gives us a remarkable opportunity to study how life may have evolved on Earth and how it may arise elsewhere in the Cosmos.

A spacecraft named *Galileo* is scheduled to reach Jupiter in 1989. When it slips into orbit around the giant planet, the spacecraft will eject an instrument-laden probe that will descend deep into Jupiter's dense atmosphere. The probe will be capable of reporting on the presence of molecules of the type that form the basic foundations of living matter. It may determine whether indeed some form of life is likely to exist on Jupiter.

An attempt to find life elsewhere in the solar system was a feature of the mission to Saturn by the *Voyager* spacecraft. After *Voyagers 1* and *2* had flown by Jupiter in

1979, they went on to Saturn, arriving there in 1980 and 1981. *Voyager 1* was guided to the vicinity of Saturn's largest moon, Titan, to search for evidence of life. Titan, the second largest moon in the whole solar system, is the only satellite known to have an atmosphere. In fact, Titan's atmosphere was so thick with clouds and haze that it interfered with all efforts to photograph the moon.

Voyager's ingenious instruments, however, revealed a great deal of information about this unusual moon. Titan's atmosphere, like that of the Earth, turns out to be composed mainly of nitrogen. It is also rich in other gases favorable to the development of amino acids and nucleotides. Astronomers suspect, too, that Titan has a solid surface where pools of liquid methane may exist. These pools could hasten the random collisions that lead to the evolution of living matter. All this might make Titan a likely place for life to begin, were it not for one discouraging factor. The atmosphere of Titan turns out to be inhospitably cold—it is minus 292 degrees Fahrenheit. Life is unlikely to arise in such a frigid environment.

Saturn is the sixth planet from the sun. It is also the second largest planet in the solar system. Saturn is 95 times more massive than the Earth. Like Jupiter, it is composed mainly of the light gases, hydrogen and helium. But Saturn lacks the striking multicolored bands of Jupiter; it has a rather bland, caramel-colored cloud cover. Before the *Voyager 1* mission, Saturn was believed to have a total of nine moons. But the *Voyager* spacecraft detected some 22 satellites orbiting the great planet, and there may be more.

The most outstanding feature of Saturn is its prominent system of rings. These extraordinary, disclike belts

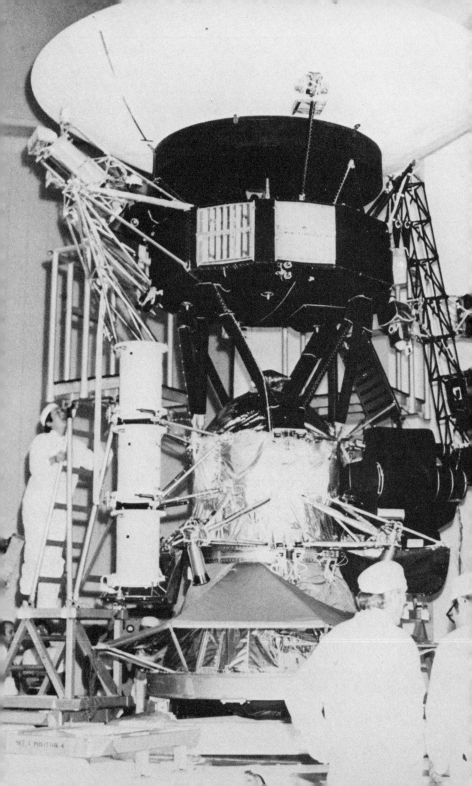

that catch and reflect the rays of the sun give the planet a special kind of symmetry and beauty. Ever since Galileo first saw them, Saturn's rings have been admired by generations of observers. The cameras of *Voyager 2* have added to our sense of wonder at these marvels.

In a series of spectacular pictures, the rings of Saturn have been shown to be far more numerous and much more complex in structure than we previously suspected. It turns out that there are actually seven major ring systems surrounding the planet, not merely the bright three we see through telescopes on Earth. Each major ring consists of hundreds of individual ringlets separated by narrow spaces somewhat like the ridges and grooves on a phonograph record. The rings are composed of countless icy particles, some tiny, others as huge as buildings. Each of these fragments is an individual satellite of Saturn whose position is determined by a balance of gravitational forces. Mysteriously, one ring has curious spokelike features, while another has two intertwined strands that appear braided.

THE VOYAGER SPACECRAFT. A Voyager spacecraft is prepared for testing by technicians at NASA's Jet Propulsion Laboratory. Seventeen feet high, with an antenna 12 feet in diameter, the complete spacecraft weighs nearly a ton. Unlike the spacecraft that visited the inner planets earlier, and depended on solar energy to power them, the Voyager craft were designed to travel beyond the range where solar panels are effective. The Voyager spacecraft, therefore, were outfitted with their own nuclear power systems. The three cylindrical units on the left side of the spacecraft shown are its nuclear generators. On the far right is a 7½-foot boom bearing the spacecraft's scientific sensing devices and two television cameras; in flight this boom extended outward. (National Aeronautics and Space Administration)

With the completion of the *Voyager* missions to Saturn, all the planets known to observers of old as the "wandering stars"—Mercury, Venus, Mars, Jupiter, and Saturn—have been explored by spacecraft. This is an achievement far beyond the greatest dreams of the high priests of Babylonia, the astrologers who studied the heavens almost 4000 years ago searching for a link between the movements of these "wandering stars" and the fate of people here on Earth.

Voyager 1 is now on its way out beyond the boundaries of the solar system into distant space. *Voyager 2* is on a course that should take it to far-off Uranus, the seventh planet from the sun, in January, 1986. Meanwhile, in 1985, NASA plans to launch a 12-ton, 94-inch telescope into Earth orbit aboard the space shuttle, *Columbia*. One of the first tasks of the space telescope will be to help select sites on Uranus for *Voyager 2* to photograph.

Astronomers know little of Uranus except that, for some reason, Uranus and its twin planet, Neptune, seem to have lost much of the lighter gases, hydrogen and helium, of which the other giant outer planets are mainly composed. Uranus appears blue-green in color through Earth-based telescopes. It has nine narrow rings and at least five moons. But the most unusual feature of Uranus is the fact

---

SATURN. *This photograph of Saturn was taken by NASA's Voyager 2 on July 21, 1981 when the spacecraft was 21 million miles from the planet. The moons Rhea and Dione appear as dots to the south and southeast of Saturn respectively. Several dark, spokelike features are apparent in the wide B ring to the left of the planet. (National Aeronautics and Space Administration)*

that it is tilted so far over on its polar axis that the planet seems to be lying on its side while it revolves around the sun.

Voyager 2 will call at Neptune in August, 1989, after visiting Uranus. Distant Neptune, the eighth planet from the sun, is nearly four times larger than Earth, and similar in size to Uranus. Neptune has two moons, and in 1982, Earth-based telescopes revealed that it may also have at least two rings. These recent observations suggest that all the giant outermost planets may be ringed. When Voyager 2 visits Neptune, it will confirm whether indeed the planet does have a ring system. Beyond Neptune, Voyager 2, like its sister ship, Voyager 1, will sail out beyond our solar system.

When the Voyager spacecraft complete their investigations of the distant planets in 1989, they will travel on beyond the limits of our solar system to begin a new mission. This mission will be a journey to the stars.

Like explorer ships of old, the Voyagers will venture out into the vast unknown. Theirs, however, may be a journey without end. It will take the spacecraft perhaps 100,000 years to reach even the nearest star, and longer still to approach others. But just maybe, at a time in the remote future, intelligent beings from some New World elsewhere in the Cosmos will find the spacecraft. For it is not likely that Earth is the only place in all the Cosmos to have life on it. The forces that created our solar system must not be unique. Other solar systems with star-orbiting planets almost certainly exist. We are surely not alone.

In hopes that the Voyagers may someday be found, they carry evidence of our existence here on Earth. There are pictures of people at their daily activities. Tokens of our civilization—recorded greetings in many languages and music from our different cultures—are aboard. So are the natural sounds of our planet, such as the songs of humpback whales. These messages of good will and gifts will outlast any monument or time capsule here on Earth. For a billion years, until the stardust of outer space finally wears them away, these symbols of our friendship will endure.

Glossary

AMINO ACIDS   Molecules that serve as the building blocks of living matter; proteins are made up of chains of these molecules.

ASTEROIDS   Small solid bodies whose orbits around the sun are typically between Mars and Jupiter.

ASTROPHYSICIST   A scientist who studies the physical properties and interactions of the stars and all other heavenly bodies.

ATOM   A tiny object that has a central core, or nucleus, and one or more orbiting electrons. The number of protons in the nucleus of the atom determines which element that particular atom represents.

ATOMIC NUMBER   The number of positive charges or protons in the nucleus of an atom of a particular element.

BIG BANG THEORY   An explanation of the origin of the Cosmos in which the entire Universe originated in one hot, explosive instant—termed the Big Bang.

BLACK HOLE   An object in space whose surface gravity is so great that nothing, not even light, can escape.

COMET   A small, icy, dusty object that becomes visible as a glowing, diffuse body with a long tail during its near approaches to the sun.

CONDENSATION   The act by which objects are reduced in volume and made more dense, as the compression of a gas to a liquid or the gathering of scattered particles into a dense solid.

CONSTELLATION   An apparent pattern of stars in the sky.

CORE (Earth's)   The central, innermost region of the Earth composed of molten nickel and iron.

CORE (star's)   The central, innermost region of a star where nuclear burning takes place.

COSMOLOGY   The study of the Universe as a whole—its past, present, and future.

COSMOS   The Universe. The totality of matter, energy, and space.

DISH, RADIO (Dish-shaped antenna)   A large, bowl-shaped surface that reflects radio waves and focuses them at the point of a radio wave detector, or antenna.

ELECTROMAGNETIC RADIATION   Waves produced by vibrating charges that travel through a vacuum at the speed of light.

ELECTROMAGNETIC SPECTRUM   The distribution of electromagnetic radiation over all possible wavelengths.

ELECTRON   A small, negatively charged particle that may normally be found in orbit around the nucleus of an atom.

ELEMENT   The identification of the type of atom based on the number of protons in the nucleus.

ELEMENTS, HEAVY   In astronomy this term refers to all elements beyond hydrogen and helium.

ELLIPSE   A geometrical figure that appears as a somewhat flattened circle. As the planets revolve about the sun, they travel in paths that are elliptical.

EXTRAGALACTIC   Outside the boundaries of a galaxy.

FREQUENCY   The number of waves, or vibrations, in a certain time interval; the number of waves, or vibrations, per second.

FUSION   The joining of two atomic nuclei with the release of energy; the basis for the hydrogen bomb in which two hydrogen nuclei fuse to form helium.

GALACTIC DISC   The highly flattened, or pancakelike, distribution of stars in a galaxy caused by the galaxy's spinning motion.

GALAXIES, CLUSTER OF   A physical clumping of galaxies that may contain a few to a few thousand galaxies.

GALAXY   A collection of stars (and often gas and dust) held together by mutual gravitation. While the actual number of stars may vary over a wide range, 100 billion is a typical number.

GALAXY, BARRED SPIRAL   A galaxy with an apparent bar running through the center.

GALAXY, ELLIPTICAL   A galaxy whose outline forms the shape of an ellipse.

GAMMA RAYS   The most energetic form of electromagnetic radiation; their wavelengths are shorter than those of x-rays.

GIANT PLANETS (also called Jovian planets) Jupiter, Saturn, Uranus, and Neptune; planets composed mainly of gases, in contrast to the smaller, solid, earthlike planets.

GRAVITY The attraction of one mass to another; one of the basic forces in the physical Universe.

GREENHOUSE EFFECT A planetary atmosphere that admits incoming solar radiation and blocks outgoing infrared (heat) radiation is said to exhibit the greenhouse effect. By sealing in the infrared radiation, the atmosphere acts like a blanket to increase the surface temperature of the planet.

HELIUM An element having two protons in its nucleus.

HELIUM BURNING The fusion process that converts helium into heavier elements with the release of enegry.

HUBBLE'S LAW The relation between the distance of an astronomical object and its red shift due to the expansion of the Cosmos.

HYDROGEN An atom with only one proton in the nucleus; the simplest element.

HYDROGEN BURNING The fusion of hydrogen to produce helium with the release of energy.

IMAGE INTENSIFIER A device similar to a television camera that strengthens the images obtained through a telescope.

INTERGALACTIC The space between galaxies.

INTERSTELLAR The space between stars.

LIGHT The visible portion of the electromagnetic spectrum.

LIGHT-YEAR　The distance that light travels at about the speed of 186,000 miles per second in one year, approximately 6 trillion (6,000,000,000,000) miles.

LOCAL GROUP　The cluster of galaxies that contains the Milky Way Galaxy.

METEORITE　Interplanetary material that has fallen to the Earth's surface (or to the surface of other planets and moons). Meteorites are responsible for the extensive cratering observed on the moon and the inner planets.

MICROWAVES　Electromagnetic waves with wavelengths near 1 centimeter (not quite half an inch).

MILKY WAY　This is the name of our own Galaxy.

MOLECULE　Two or more atoms that are bound together.

NEBULA　A term given to any diffuse, or cloudlike, astronomical object.

NEUTRON　A neutral atomic particle that is approximately 2000 times more massive than an electron. The neutron has almost the same mass as a proton and is found in the nuclei of all atoms except hydrogen.

NEUTRON STAR　A star of extremely high density that is composed primarily of neutrons. The mass may be a few times the mass of our sun, but the object is extremely compacted with a radius of merely 10 miles.

NUCLEAR BURNING　The process of fusion of atomic nuclei with the release of energy.

NUCLEAR ENERGY　Energy derived from the fusion or fission (splitting) of the nucleus of an atom.

NUCLEAR REACTION　A process of either fusion or fission in the nucleus.

NUCLEOTIDES Molecules formed by joined chains of amino acids. Nucleotides combine to form complex molecules, nucleic acids, capable of reproducing themselves (for example, DNA).

NUCLEUS (atomic) The central region of an atom usually containing protons and neutrons in nearly equal numbers.

NUCLEUS (of a galaxy) The central region of a galaxy that contains the greatest density of stars.

OPTICAL TELESCOPE An astronomical instrument that concentrates visible light, an aid for observing distant objects, such as the stars and planets.

OSCILLATING THEORY A concept that combines the Big Bang cosmology with the notion that the Universe undergoes repeated cycles of expansion, collapse, and rebirth.

PLANETESIMAL A small object circling the sun like a miniature planet.

PRISM A wedge-shaped, solid piece of glass that disperses light to produce a spectrum.

PROTON A positively charged particle found in the nucleus of an atom.

PULSAR A rapidly rotating neutron star. Active areas on the surface of the neutron star emit streams of radiation into space. Because of the rotation, a stream sweeps across the observer's line of sight at regularly timed intervals, creating a succession of pulses, or flashes of radiation as seen from the Earth.

QUASARS   Objects that resemble stars but have energy outputs generally exceeding those of galaxies. These luminous bodies are the brightest and most distant objects known in the Universe.

RADAR (*radio detecting and ranging*)   A device used to detect the presence and distance of an object by measuring the time it takes for a beamed radio wave to bounce back from the object's surface. Radar is used to determine the direction and speed of objects such as airplanes, ships, automobiles, and weather fronts.

RADIATION, INFRARED   Electromagnetic radiation of a wavelength just longer than visible radiation.

RADIO ASTRONOMY   The field of astronomy concerned with observation and interpretation of radio waves from space.

RADIO GALAXY   A galaxy that is a strong source of radio waves.

RADIO TELESCOPE   A telescope used for focusing and recording radio waves.

RADIO WAVES   Electromagnetic waves of long wavelength, typically several meters or more.

RADIOACTIVE   An element is said to be radioactive if its nucleus undergoes spontaneous disintegration emitting radiant energy in the form of particles or rays. Elements such as thorium, uranium, plutonium, and radium are radioactive.

RED GIANT   A mature star near the end of its hydrogen-burning stage showing signs of age, a swelling and reddening of its outer layers. A red giant is a star that has become a huge red ball typically 100 times larger than its original size and much brighter.

RED SHIFT   A shift toward the longer wavelengths at the red end of the visible light spectrum that is exhibited by retreating galaxies.

REFLECTING TELESCOPE   An optical telescope that employs a concave, reflecting mirror to concentrate light.

REFRACTING TELESCOPE   An optical telescope that uses a lens, or system of lenses, to concentrate light.

SATELLITE   An object that is in orbit around another object, usually a planet. Moons are natural satellites; man-made satellites, boosted into orbit by rockets, serve a variety of scientific, communications and military purposes.

SEISMOMETER   The detecting portion of a seismograph, an instrument that records earthquakes.

SOLAR SYSTEM   The sun and all the objects orbiting it: planets, moons, asteroids, and comets.

SPECTRUM   The array of colors (visible spectrum) or intensities of radiation (electromagnetic spectrum) at different wavelengths presented in order of their wavelengths.

SPIRAL ARM   A curved luminous arm extending from the center of a spiral galaxy.

STAR   A gaseous sphere held together by its own gravity; especially, one massive enough to provide internal temperatures needed for nuclear burning.

STEADY-STATE THEORY   A concept that holds that the overall properties of the Cosmos do not change with the passage of time. This cosmology implies that a constant creation of new matter and energy keeps the Cosmos in balance.

SUPERNOVA The final catastrophic end of a large, massive star. When nuclear burning is all but complete and the pressure at the center of the star is sufficiently great, the dying star explodes leaving nothing behind but the squeezed remnant of its original core. The explosion hurls out into space all the elements produced during the lifetime of the star.

ULTRAVIOLET LIGHT Light whose wavelength is just shorter than that of visible light.

VISIBLE LIGHT Electromagnetic radiation which the eye can see.

VISIBLE SPECTRUM Observable light arranged in order of wavelength from short to long: violet, blue, green, yellow, orange, red.

WAVELENGTH The distance between two successive peaks or valleys in a wave.

WHITE DWARF A small hot star that is near or at the end of its nuclear burning phase.

X-RAYS Radiation with wavelengths shorter than those of ultraviolet light.

Index